# Studies in Fuzziness and Soft Computing 295

**Editor-in-Chief**

Prof. Janusz Kacprzyk
Systems Research Institute
Polish Academy of Sciences
ul. Newelska 6
01-447 Warsaw
Poland
E-mail: kacprzyk@ibspan.waw.pl

T0140423

For further volumes:
http://www.springer.com/series/2941

Barnabas Bede

# Mathematics of Fuzzy Sets and Fuzzy Logic

 Springer

*Author*
Barnabas Bede
Department of Mathematics
DigiPen Institute of Technology
Redmond, WA
USA

ISSN 1434-9922                         e-ISSN 1860-0808
ISBN 978-3-642-43302-3                 ISBN 978-3-642-35221-8  (eBook)
DOI 10.1007/978-3-642-35221-8
Springer Heidelberg New York Dordrecht London

Printed on acid-free paper

Springer is part of Springer Science+Business Media (www.springer.com)

*Dedicated to my daughters: Fruzsina and Tekla*

# Preface

The present manuscript is intended to be a textbook that could serve both undergraduate and graduate students when studying Fuzzy Set Theory and Applications. It is also intended to deepen the research into some existing directions and to investigate some new research directions. The book tries to develop a systematic, Mathematically-based introduction into the theory and applications of fuzzy sets and fuzzy logic. In this way the author tries to cover a gap in the literature. Also, the book can be an introduction into Fuzzy Sets and Systems for researchers who are interested in the topic of fuzzy sets in all areas of Mathematics, Computer Science and Engineering, or simply interested Mathematicians, Engineers and students in these areas. The book starts from the basic theory and gets the reader to a level very close to the current research topics in Fuzzy Sets.

The basis of the book is the author's class notes for Fuzzy Sets and Fuzzy Logic, class taught at DigiPen Institute of Technology.

Another goal that the author had when writing the present book is try to see where Fuzzy Sets and Fuzzy Logic as a discipline, can be connected to other areas of Mathematics. The manuscript tries to show that Fuzzy Sets is an independent discipline having huge overlaps on one side with Analysis and Approximation Theory and on the other side with Logic and Set Theory. The approach of the present work leans toward Mathematical Analysis and Approximation.

I am really thankful to my wife Emese, who provided very strong support for me in writing this book. Also, the author is thankful to his colleagues, especially Sorin G. Gal, L. Stefanini, L.C. Barros, I.J. Rudas and many others for comments that improved the manuscript. The author would like to express his thanks to all the students who contributed with comments that improved the manuscript, especially Kia McDowel, Matt Peterson, Chris Barrett, and many others.

Barnabas Bede
Redmond, WA

# Contents

# 1
# Fuzzy Sets

## 1.1 Classical Sets

The concept of a **set** is fundamental in Mathematics and intuitively can be described as a collection of objects possibly linked through some properties. A classical set has clear boundaries, i.e. $x \in A$ or $x \notin A$ exclude any other possibility.

**Definition 1.1.** *Let $X$ be a set and $A$ be a subset of $X$ ($A \subseteq X$). Then the function*

$$\chi_A(x) = \begin{cases} 1 & if \quad x \in A \\ 0 & if \quad x \notin A \end{cases}$$

*is called the characteristic function of the set $A$ in $X$.*

Classical sets and their operations can be represented by their characteristic functions.

Indeed, let us consider the union $A \cup B = \{x \in X | x \in A \text{ or } x \in B\}$. Its characteristic function is

$$\chi_{A \cup B}(x) = \max\{\chi_A(x), \chi_B(x)\}.$$

For the intersection $A \cap B = \{x \in X | x \in A \text{ and } x \in B\}$ the characteristic function is

$$\chi_{A \cap B}(x) = \min\{\chi_A(x), \chi_B(x)\}.$$

If we consider the complement of $A$ in $X$, $\bar{A} = \{x \in X | x \notin A\}$ it has the characteristic function

$$\chi_{\bar{A}}(x) = 1 - \chi_A(x).$$

B. Bede: *Mathematics of Fuzzy Sets and Fuzzy Logic*, STUDFUZZ 295, pp. 1–12.
DOI: 10.1007/978-3-642-35221-8_1    © Springer-Verlag Berlin Heidelberg 2013

## 1.2   Fuzzy Sets

Fuzzy sets were introduced by L. Zadeh in [154]. The definition of a fuzzy set given by L. Zadeh is as follows: A **fuzzy set** is a class with a continuum of membership grades. So a fuzzy set $A$ in a referential (universe of discourse) $X$ is characterized by a membership function $A$ which associates with each element $x \in X$ a real number $A(x) \in [0, 1]$, having the interpretation $A(x)$ is the membership grade of $x$ in the fuzzy set $A$.

**Definition 1.2.** *(Zadeh [154]) A fuzzy set $A$ (fuzzy subset of $X$) is defined as a mapping*

$$A : X \to [0, 1],$$

*where $A(x)$ is the membership degree of $x$ to the fuzzy set $A$. We denote by $\mathcal{F}(X)$ the collection of all fuzzy subsets of $X$.*

Fuzzy sets are generalizations of the classical sets represented by their characteristic functions $\chi_A : X \to \{0, 1\}$. In our case $A(x) = 1$ means full membership of $x$ in $A$, while $A(x) = 0$ expresses non-membership, but in contrary to the classical case other membership degrees are allowed.

We identify a fuzzy set with its membership function. Other notations that can be used are the following $\mu_A(x) = A(x)$.

Every classical set is also a fuzzy set. We can define the membership function of a classical set $A \subseteq X$ as its characteristic function

$$\mu_A(x) = \begin{cases} 1 & \text{if} \quad x \in A \\ 0 & \text{otherwise} \end{cases} .$$

Fuzzy sets are able to model linguistic uncertainty and the following examples show how:

**Example 1.3.** *In this example we consider the expression "young" in the context "a young person" in order to exemplify how linguistic expressions can be modeled using fuzzy sets. The fuzzy set $A : [0, 100] \to [0, 1]$,*

$$A(x) = \begin{cases} 1 & \text{if} \quad 0 \le x \le 20 \\ \frac{40-x}{20} & \text{if} \quad 20 < x \le 40 \\ 0 & \text{otherwise} \end{cases}$$

*is illustrated in Fig. 1.1.*

**Example 1.4.** *Let us consider the fuzzy set $A : \mathbb{R} \to [0, 1]$, $A(x) = \frac{1}{1+x^2}$. This fuzzy set can model the linguistic expression "real number near 0" (see Fig. 1.2).*

**Example 1.5.** *If given a crisp parameter which is known only through an expert's knowledge and we know that its values are in the $[0, 60]$ interval, the expert's knowledge expressed in terms of estimates small, medium, and high can be modeled, e.g. by the fuzzy sets in Fig. 1.3.*

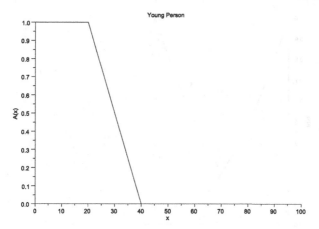

**Fig. 1.1** Example of a fuzzy set for modeling the expression young person

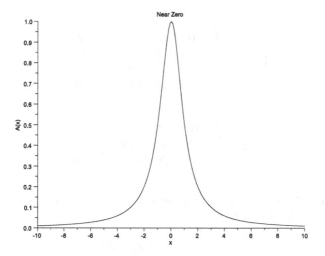

**Fig. 1.2** Fuzzy set that models a real number near 0

**Example 1.6.** *Fuzzy sets can be used to express subjective perceptions in a mathematical form. Let $X = [40, 100]$ be the interval of temperatures for a room. Fuzzy sets $A_1, A_2, ..., A_5$ can be used to model the perceptions: cold, cool, just right, warm, and hot (see Figure 1.4):*
*cold:*

$$A_1(x) = \begin{cases} 1 & if \quad 40 \leq x < 50 \\ \frac{60-x}{10} & if \quad 50 \leq x < 60 \\ 0 & if \quad 60 \leq x \leq 100 \end{cases}$$

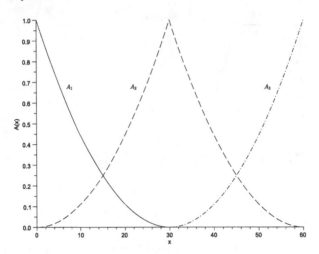

**Fig. 1.3** Expert knowledge represented by fuzzy sets

*cool:*

$$A_2(x) = \begin{cases} 0 & if \quad 40 \leq x < 50 \\ \frac{x-50}{10} & if \quad 50 \leq x < 60 \\ \frac{70-x}{10} & if \quad 60 \leq x < 70 \\ 0 & if \quad 70 \leq x \leq 100 \end{cases}$$

...

*hot:*

$$A_5(x) = \begin{cases} 0 & if \quad 40 \leq x < 80 \\ \frac{x-80}{10} & if \quad 80 \leq x < 90 \\ 1 & if \quad 90 \leq x \leq 100 \end{cases} \quad .$$

**Definition 1.7.** *Let $A : X \to [0,1]$ be a fuzzy set. The **level sets** of $A$ are defined as the classical sets*

$$A_\alpha = \{x \in X | A(x) \geq \alpha\},$$

$0 < \alpha \leq 1.$

$$A_1 = \{x \in X | A(x) \geq 1\}$$

*is called the **core** of the fuzzy set $A$, while*

$$\operatorname{supp} A = \{x \in X | A(x) > 0\}$$

*is called the **support** of the fuzzy set.*

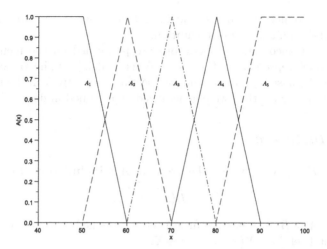

**Fig. 1.4** Fuzzy sets Cold, Cool, Just Right, Warm and Hot used in a room temperature control example

**Example 1.8.** *Let us consider the cool fuzzy set as in the previous example.*

$$A_2(x) = \begin{cases} 0 & \text{if,} & 40 \le x < 50 \\ \frac{x-50}{10} & \text{if} & 50 \le x < 60 \\ 1 - \frac{x-60}{10} & \text{if} & 60 \le x < 70 \\ 0 & \text{if} & 70 \le x \le 100 \end{cases}.$$

*Its core is* $(A_2)_1 = \{60\}$, *the* $\frac{1}{2}$-*level set is* $(A_2)_{\frac{1}{2}} = [55, 65]$, *the* $\alpha$-*level set is* $(A_2)_\alpha = [50+10\alpha, 70-10\alpha]$, $0 < \alpha \le 1$ *and the support is* $\text{supp } A_2 = (50, 70)$.

**Remark 1.9.** *If the universe of discourse is a finite set* $X = \{x_1, x_2, ..., x_n\}$ *then a fuzzy set* $A : X \to [0, 1]$ *can be represented formally as*

$$A = \frac{A(x_1)}{x_1} + \frac{A(x_2)}{x_2} + ... + \frac{A(x_n)}{x_n}.$$

**Example 1.10.** *Let us consider the expression "good grade in Mathematics".* *This expression can be represented as a fuzzy set* $G : \{A, B, C, D, F\} \to [0, 1]$, $G = \frac{1}{A} + \frac{0.7}{B} + \frac{0.3}{C} + \frac{0}{D} + \frac{0}{F}$. *The core of* $G$ *is* $G_1 = \{A\}$, *the support is* $\text{supp } G = \{A, B, C\}$ *and the* $\frac{1}{2}-$ *level set is* $G_{\frac{1}{2}} = \{A, B\}$.

## 1.3 The Basic Connectives

Let $\mathcal{F}(X)$ denote the collection of fuzzy sets on a given universe of discourse $X$.

The basic connectives in fuzzy logic and fuzzy set theory are inclusion, union, intersection and complementation.

In fuzzy set theory these operations are performed on the membership functions which represent the fuzzy sets. When Zadeh [154], introduced these connectives, he based the union and intersection connectives on the max and min operations. Later they were generalized and studied in detail.

### 1.3.1   Inclusion

Let $A, B \in \mathcal{F}(X)$. We say that the fuzzy set $A$ is **included** in $B$ if

$$A(x) \leq B(x), \forall x \in X.$$

We denote $A \leq B$. The empty (fuzzy) set $\emptyset$ is defined as $\emptyset(x) = 0, \forall x \in X$, and the total set $X$ is $X(x) = 1, \forall x \in X$.

### 1.3.2   Intersection

Let $A, B \in \mathcal{F}(X)$. The **intersection** of $A$ and $B$ is the fuzzy set $C$ with

$$C(x) = \min\{A(x), B(x)\} = A(x) \wedge B(x), \forall x \in X.$$

We denote $C = A \wedge B$.

### 1.3.3   Union

Let $A, B \in \mathcal{F}(X)$. The **union** of $A$ and $B$ is the fuzzy set $C$, where

$$C(x) = \max\{A(x), B(x)\} = A(x) \vee B(x), \forall x \in X.$$

We denote $C = A \vee B$.

### 1.3.4   Complementation

Let $A \in \mathcal{F}(X)$ be a fuzzy set. The **complement** of $A$ is the fuzzy set $B$ where

$$B(x) = 1 - A(x), \forall x \in X.$$

We denote $B = \bar{A}$.

**Remark 1.11.** *We observe that the operations between fuzzy sets are defined point-wise in terms of operations on the $[0, 1]$ interval. Also, let us mention here that throughout the text we will use the following notations $\min\{x, y\} = x \wedge y$, $\max\{x, y\} = x \vee y$ with operands $x, y \in [0, 1]$ or $x, y \in \mathbb{R}$. Also, without the danger of a major confusion we use the notations $A \wedge B$ and $A \vee B$ to denote the intersection and union of two fuzzy sets.*

**Example 1.12.** *If we consider the fuzzy sets*

$$A_1(x) = \begin{cases} 1 & if \quad 40 \le x < 50 \\ 1 - \frac{x-50}{10} & if \quad 50 \le x < 60 \\ 0 & if \quad 60 \le x \le 100 \end{cases},$$

$$A_2(x) = \begin{cases} 0 & if \quad 40 \le x < 50 \\ \frac{x-50}{10} & if \quad 50 \le x < 60 \\ 1 - \frac{x-60}{10} & if \quad 60 \le x < 70 \\ 0 & if \quad 70 \le x \le 100 \end{cases}$$

*given in Example 1.6, then their union is*

$$A_1 \vee A_2(x) = \begin{cases} 1 & if \quad 40 \le x < 50 \\ 1 - \frac{x-50}{10} & if \quad 50 \le x < 55 \\ \frac{x-50}{10} & if \quad 55 \le x \le 60 \\ 1 - \frac{x-60}{10} & if \quad 60 \le x \le 70 \\ 0 & if \quad 70 < x \le 100 \end{cases}.$$

*The intersection can be expressed as*

$$A_1 \wedge A_2(x) = \begin{cases} 0 & if \quad 40 \le x < 50 \\ \frac{x-50}{10} & if \quad 50 \le x < 55 \\ 1 - \frac{x-50}{10} & if \quad 55 \le x \le 60 \\ 0 & if \quad 60 < x \le 100 \end{cases}.$$

*The complement of $A_1$ can be written*

$$\bar{A}_1(x) = \begin{cases} 0 & if \quad 40 \le x < 50 \\ \frac{x-50}{10} & if \quad 50 \le x < 60 \\ 1 & if \quad 60 \le x \le 100 \end{cases}$$

*see Figs. 1.5, 1.6, 1.7.*

**Example 1.13.** *If the universe of discourse is a discrete set $X = \{x_1, x_2, ..., x_n\}$ then the union intersection and complementation can be easily expressed. If $A, B : X \to [0,1]$, they can be represented as*

$$A = \frac{A(x_1)}{x_1} + \frac{A(x_2)}{x_2} + ... + \frac{A(x_n)}{x_n}.$$

$$B = \frac{B(x_1)}{x_1} + \frac{B(x_2)}{x_2} + ... + \frac{B(x_n)}{x_n}.$$

*Then the union is*

$$A \vee B = \frac{A(x_1) \vee B(x_1)}{x_1} + \frac{A(x_2) \vee B(x_2)}{x_2} + ... + \frac{A(x_n) \vee B(x_n)}{x_n},$$

*the intersection is*

$$A \wedge B = \frac{A(x_1) \wedge B(x_1)}{x_1} + \frac{A(x_2) \wedge B(x_2)}{x_2} + ... + \frac{A(x_n) \wedge B(x_n)}{x_n}$$

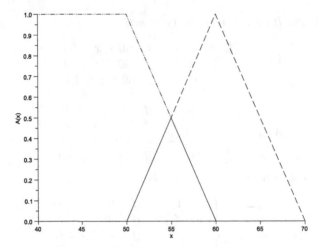

**Fig. 1.5** Fuzzy Intersection

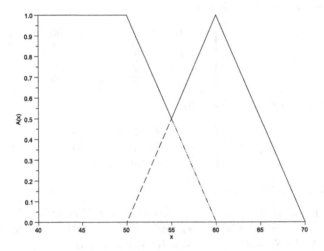

**Fig. 1.6** Union of two fuzzy sets

*and the complement of A is*

$$\bar{A} = \frac{1 - A(x_1)}{x_1} + \frac{1 - A(x_2)}{x_2} + ... + \frac{1 - A(x_n)}{x_n}.$$

**Example 1.14.** *Consider $X = [0, \infty]$ and the fuzzy sets $A(x) = \frac{x}{x+1}$, $B(x) = \frac{1}{x^2+1}$. Then we can illustrate $A \vee B$, $A \wedge B$ and $\bar{A}$ as in figs. 1.8, 1.9.*

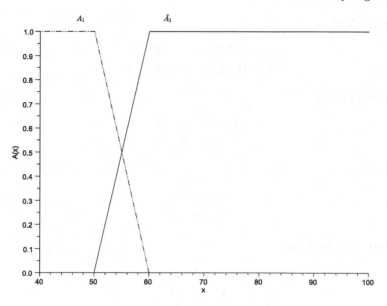

**Fig. 1.7** The complement of a fuzzy set

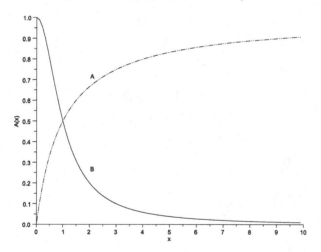

**Fig. 1.8** Two fuzzy sets $A$ and $B$

## 1.4 Fuzzy Logic

**Proposition 1.15.** *(see e.g., Dubois-Prade [51]) Considering the basic connectives in fuzzy set theory, the following properties hold true:*

*1. Associativity*

$$A \wedge (B \wedge C) = (A \wedge B) \wedge C$$
$$A \vee (B \vee C) = (A \vee B) \vee C$$

*2. Commutativity*

$$A \wedge B = B \wedge A$$
$$A \vee B = B \vee A$$

*3. Identity*

$$A \wedge X = A$$
$$A \vee \emptyset = A$$

*4. Absorption by $\emptyset$ and $X$*

$$A \wedge \emptyset = \emptyset$$
$$A \vee X = X$$

*5. Idempotence*

$$A \wedge A = A$$
$$A \vee A = A$$

*6. De Morgan Laws*

$$\overline{A \wedge B} = \bar{A} \vee \bar{B}$$
$$\overline{A \vee B} = \bar{A} \wedge \bar{B}$$

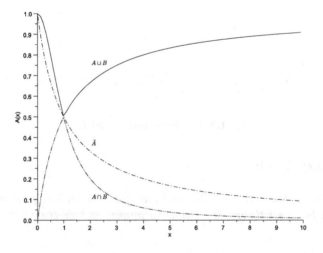

**Fig. 1.9** Basic connectives for the fuzzy sets $A$ and $B$

*7. Distributivity*

$$A \wedge (B \vee C) = (A \wedge B) \vee (A \wedge C)$$
$$A \vee (B \wedge C) = (A \vee B) \wedge (A \vee C)$$

*8. Involution*

$$\overline{\overline{A}} = A$$

*9. Absorption*

$$A \wedge (A \vee B) = A$$
$$A \vee (A \wedge B) = A$$

**Proof.** The proofs of properties 1-5, 7, and 8 are left to the reader.

Let us prove the first De Morgan law in 6. Let $x \in X$. Then

$$\overline{A \wedge B}(x) = 1 - \min\{A(x), B(x)\}$$
$$= \max\{1 - A(x), 1 - B(x)\} = \bar{A} \vee \bar{B}(x).$$

Let us also prove one of the absorption laws in 9: $A \vee (A \wedge B) = A$. Let $x \in X$. Then we have

$$A \vee (A \wedge B)(x) = A(x) \vee (A(x) \wedge B(x)) \leq A(x) \vee A(x)$$
$$= A(x) \leq A(x) \vee (A(x) \wedge B(x)).$$

■

The algebraic structure obtained in this way is called a distributive pseudo-complemented lattice.

Let us remark that the laws of contradiction and excluded middle ("tertio non datur") fail. More precisely:

**Proposition 1.16.** *If A is a non-classical fuzzy set $A : X \to [0,1]$ (i.e., there exists $x \in X$ with $A(x) \notin \{0,1\}$) then*

$$A \wedge \bar{A} \neq \emptyset$$
$$A \vee \bar{A} \neq X.$$

**Proof.** If $x \in X$ is such that $0 < A(x) < 1$ then $0 < \bar{A}(x) < 1$ and then $0 < A \wedge \bar{A}(x) < 1$ and $0 < A \vee \bar{A}(x) < 1$.   ■

So, if we allow gradual membership for fuzzy sets, then the algebraic structure is incompatible with the Boolean algebra structure which is at the basis of classical set theory and classical logic. As a conclusion, we need a different theory. This theory is the theory of fuzzy sets and fuzzy logic.

## 1.5   Problems

1. Set up membership functions for modeling linguistic expressions about the speed of a car on a highway: "very slow", "slow", "average", "fast", "very fast".

2. Consider the fuzzy sets $A, B : \{1, 2, ..., 10\} \to [0, 1]$ defined as $A(x) = \frac{0}{1} + \frac{0.2}{2} + \frac{0.7}{3} + \frac{1}{4} + \frac{0.7}{5} + \frac{0.2}{6} + \frac{0}{7} + \frac{0}{8} + \frac{0}{9} + \frac{0}{10}$, $B(x) = \frac{0}{1} + \frac{0}{2} + \frac{0}{3} + \frac{0.3}{4} + \frac{0.5}{5} + \frac{0.8}{6} + \frac{1}{7} + \frac{0.5}{8} + \frac{0.2}{9} + \frac{0}{10}$,. Calculate $A \wedge B$, $A \vee B$, $\bar{A}$, $\bar{B}$. Calculate and compare $\bar{A} \vee \bar{B}$ and $\overline{A \wedge B}$.

3. Consider the fuzzy sets $A, B : \mathbb{R}_+ \to [0, 1]$ defined as $A(x) = \frac{1}{1+x^2}$, $B(x) = \frac{1}{10^x}$. Calculate $A \wedge B$, $A \vee B$, $\bar{A}$, $\bar{B}$ and graph them.

4. Consider the fuzzy sets

$$
A(x) = \begin{cases} 0 & \text{if,} & x < 1 \\ \frac{x-1}{6} & \text{if} & 1 \leq x < 7 \\ \frac{10-x}{3} & \text{if} & 7 \leq x < 10 \\ 0 & \text{if} & 10 \leq x \end{cases} ,
$$

$$
B(x) = \begin{cases} 0 & \text{if,} & x < 2 \\ x - 2 & \text{if} & 2 \leq x < 3 \\ 1 & \text{if} & 3 \leq x < 4 \\ \frac{6-x}{2} & \text{if} & 4 \leq x \leq 6 \\ 0 & \text{if} & 6 < x \end{cases} .
$$

Find $A \wedge B$, $A \vee B$, $\bar{A}$ and graph them.

5. Prove the properties 1-5, 7,8 in Proposition 1.15.

6. Prove that for any fuzzy set $A \in \mathcal{F}(X)$ we have

$$A \vee \bar{A} \geq 0.5,$$

while

$$A \wedge \bar{A} \leq 0.5.$$

7. Prove that for any fuzzy sets $A, B \in \mathcal{F}(X)$ we have

$$(A \wedge \bar{B}) \vee (\bar{A} \wedge B) \geq 0.5 \wedge (A \vee B) \wedge (\bar{A} \vee \bar{B})$$

and

$$(A \vee \bar{B}) \wedge (\bar{A} \vee B) \leq 0.5 \vee (\bar{A} \wedge \bar{B}) \vee (A \wedge B).$$

# 2

# Fuzzy Set-Theoretic Operations

## 2.1 Negation

As we have seen in the previous section, we can identify the operations between fuzzy sets by the corresponding operations on the unit interval, fuzzy
set operations being defined point-wise. This implies that we can study operations between fuzzy sets by the corresponding operations over the real unit
interval.

**Definition 2.1.** *(Trillas [147], see also Fodor-Roubens [60]) A function*
$N : [0,1] \rightarrow [0,1]$ *is called a **negation** if* $N(0) = 1$, $N(1) = 0$ *and* $N$ *is
non-increasing* $(x \le y \Rightarrow N(x) \ge N(y))$. *A negation is called a **strict negation** if it is strictly decreasing* $(x < y \Rightarrow N(x) > N(y))$ *and continuous. A
strict negation is said to be a **strong negation** if it is also involutive, i.e.*
$N(N(x)) = x$.

**Definition 2.2.** *If* $A \in \mathcal{F}(X)$ *is a fuzzy set then the* $N-$***complement** *of* $A$
*is defined point-wise as* $N(A)(x) = N(A(x))$.

**Example 2.3.** *The **standard negation** is* $N(x) = 1 - x$, *and it is a strong
negation.*

**Example 2.4.** *The negation* $N_\lambda(x) = \frac{1-x}{1+\lambda x}$, $\lambda > -1$, *is a strong negation
called the* $\lambda$-*complement.*

**Remark 2.5.** *An alternative notation for any negation* $N(x)$ *is* $\bar{x}$.

In the following theorem, strong negations are characterized:

B. Bede: *Mathematics of Fuzzy Sets and Fuzzy Logic*, STUDFUZZ 295, pp. 13–31.
DOI: 10.1007/978-3-642-35221-8_2     © Springer-Verlag Berlin Heidelberg 2013

**Theorem 2.6.** *(Trillas [147]) A function $N : [0,1] \to [0,1]$ is a strong negation if and only if there exists an automorphism (continuous invertible function $\varphi : [0,1] \to [0,1]$ with continuous inverse $\varphi^{-1}$) such that*

$$N(x) = \varphi^{-1}(1 - \varphi(x)).$$

**Proof.** " $\Rightarrow$ " If $n$ is a strict negation then $f(x) = N(x) - x$ is continuous, $f(0) = 1$, and $f(1) = -1$. In which case there exists an $x^* \in (0,1)$ such that $f(x^*) = 0$, or equivalently, $N(x^*) = x^*$, i.e., $N$ has a fixed point.

Now let $N_1, N_2$ be two strict negations. Then there exist $s_1, s_2 \in (0,1)$ such that $N_1(s_1) = s_1$ and $N_2(s_2) = s_2$. Let $t = \frac{s_2}{s_1}$ and $\varphi, \psi : [0,1] \to [0,1]$,

$$\varphi(x) = \begin{cases} \frac{x}{t} & \text{if } x \le s_2 \\ N_1^{-1}\left(\frac{N_2(x)}{t}\right) & \text{if } x > s_2 \end{cases},$$

$$\psi(x) = \begin{cases} tx & \text{if } x \le s_1 \\ N_2\left(tN_1^{-1}(x)\right) & \text{if } x > s_1 \end{cases}.$$

We claim that $N_2 = \psi \circ N_1 \circ \varphi$. Indeed, if $x < s_2$ then $\frac{x}{t} < \frac{s_2}{t} = s_1$, and $N_1(\frac{x}{t}) > N_1(s_1) = s_1$. We have

$$\psi \circ N_1 \circ \varphi(x) = \psi(N_1(\frac{x}{t}))$$

$$= N_2[t \cdot N_1^{-1}(N_1(\frac{x}{t}))] = N_2\left(t \cdot \frac{x}{t}\right) = N_2(x).$$

If $x \ge s_2$ then $N_2(x) \le N_2(s_2) = s_2 = s_1 t$. Then we obtain $\frac{N_2(s_2)}{t} \le s_1$. We have

$$\psi \circ N_1 \circ \varphi(x) = \psi\left(N_1\left(N_1^{-1}\left(\frac{N_2(x)}{t}\right)\right)\right)$$

$$= \psi\left(\frac{N_2(x)}{t}\right) = t \cdot \frac{N_2(x)}{t} = N_2(x).$$

As a conclusion we obtain $\psi \circ N_1 \circ \varphi = N_2$. Let $N_1$ be the standard negation $N_1(x) = 1 - x$. Then

$$\varphi(x) = \begin{cases} \frac{x}{t} & \text{if } x \le s_2 \\ 1 - \frac{N_2(x)}{t} & \text{if } x > s_2 \end{cases},$$

$$\psi(x) = \begin{cases} tx & \text{if } x \le \frac{1}{2} \\ N_2(t(1-x)) & \text{if } x > \frac{1}{2} \end{cases}$$

and $t = 2s_2$.

If $x \le s_2$ then $\frac{x}{t} \le \frac{s_2}{t} = \frac{1}{2}$, i.e., $\varphi(x) \le \frac{1}{2}$ and then we have

$$\psi \circ \varphi(x) = t\varphi(x) = t \cdot \frac{x}{t} = x.$$

If $x > s_2$ then $N_2(x) < N_2(s_2) = s_2$ and we obtain $1 - \frac{N_2(x)}{t} > 1 - \frac{s_2}{t} = \frac{1}{2}$, i.e., $\varphi(x) > \frac{1}{2}$. Taking these into account we have

$$\psi \circ \varphi(x) = N_2(t(1 - \varphi(x))) = N_2\left(t\left(\frac{N_2(x)}{t}\right)\right) = N_2(N_2(x)).$$

If $N_2$ is a strong negation then $N_2(N_2(x)) = x$, i.e., $\psi \circ \varphi(x) = x$ and as a conclusion $\psi \circ \varphi = 1_{[0,1]}$

Now, let us calculate $\varphi \circ \psi$.

If $x \leq \frac{1}{2}$ then $xt \leq \frac{t}{2} = s_2$, i.e., $\psi(x) \leq \frac{1}{2}$. Then $\varphi \circ \psi(x) = \varphi(xt) = \frac{xt}{t} = x$.

If $x > \frac{1}{2}$ then $1 - x < \frac{1}{2}$, then $t(1 - x) \leq \frac{t}{2} = s_2$ and $N_2(t(1 - x)) \geq N_2(s_2) = s_2$. Then

$$\varphi \circ \psi(x) = \varphi(N_2(t(1 - x))) = 1 - \frac{N_2(N_2(t(1 - x)))}{t}.$$

If $N_2$ is a strong negation we get

$$\varphi \circ \psi(x) = 1 - \frac{t(1 - x)}{t} = x.$$

The continuity of $\varphi$ and $\psi$ is easy to be checked and we obtain $\varphi \circ \psi = 1_{[0,1]}$, i.e., $\varphi$ and $\psi$ are inverses of each other ($\psi = \varphi^{-1}$).

As a conclusion if $N$ is a strong negation then there exists an automorphism $\varphi : [0, 1] \to [0, 1]$, such that $N(x) = \varphi^{-1}(1 - \varphi(x))$.

"$\Leftarrow$" Now given $\varphi$, an automorphism of $[0, 1]$, we consider $N(x) = \varphi^{-1}(1 - \varphi(x))$. Then $N$ is obviously continuous. Also,

$$N(0) = \varphi^{-1}(1 - \varphi(0)) = \varphi^{-1}(1) = 1$$

$$N(1) = \varphi^{-1}(1 - \varphi(1)) = \varphi^{-1}(0) = 0,$$

and

$$N(N(x)) = \varphi^{-1}(1 - \varphi(\varphi^{-1}(1 - \varphi(x))))$$

$$= \varphi^{-1}(1 - 1 + \varphi(x)) = \varphi^{-1}(\varphi(x)) = x.$$

To show that $N$ is strictly decreasing, we observe that $\varphi$ is increasing, $1 - \varphi$ is decreasing and $\varphi^{-1}$ is increasing. Then $\varphi^{-1}(1 - \varphi(x))$ is strictly decreasing. As a conclusion $N$ is a strong negation. ∎

**Remark 2.7.** $N(x) = \varphi^{-1}(1 - \varphi(x))$ *is called the $\varphi$-transform of the standard negation.*

**Example 2.8.** *A very popular parametric family of strong negations is* $N(x) = (1 - x^\alpha)^{\frac{1}{\alpha}}$, *for any* $\alpha \in (0, \infty)$.

## 2.2   Triangular Norms and Conorms

Triangular norms and conorms generalize the basic connectives between fuzzy sets. They were first introduced in the theory of probabilistic metric spaces. Later these were found to be very suitable to be used with fuzzy sets.

**Definition 2.9.** *(Schweizer-Sklar [130]) Let $T, S : [0,1]^2 \to [0,1]$. Consider the following properties:*

$T_1 : T(x,1) = x$ *(identity)*
$S_1 : S(x,0) = x$
$T_2 : T(x,y) = T(y,x)$ *(commutativity)*
$S_2 : S(x,y) = S(y,x)$
$T_3 : T(x,T(y,z)) = T(T(x,y),z)$ *(associativity)*
$S_3 : S(x,S(y,z)) = S(S(x,y),z)$
$T_4 :$ *If $x \leq u$ and $y \leq v$ then $T(x,y) \leq T(u,v)$ (monotonicity)*
$S_4 :$ *If $x \leq u$ and $y \leq v$ then $S(x,y) \leq S(u,v)$*

*A **triangular norm** (t-norm) is a function $T : [0,1]^2 \to [0,1]$ that satisfies $T_1 - T_4$.*

*A **triangular conorm** (t-conorm or s-norm) is a function $S : [0,1]^2 \to [0,1]$ that satisfies $S_1 - S_4$.*

We will occasionally use the notation $xTy = T(x,y)$, $xSy = S(x,y)$.

**Definition 2.10.** *Given $T, S : [0,1]^2 \to [0,1]$ a t-norm and a t-conorm, for fuzzy sets $A, B \in \mathcal{F}(X)$ we define operations point-wise*

$$ATB(x) = A(x)TB(x)$$
$$ASB(x) = A(x)SB(x)$$

*for any $x \in X$.*

**Proposition 2.11.** *Given any t-norm $T$ and t-conorm $S$, we have $T(x,0) = 0$ and $S(x,1) = 1$, for all $x \in [0,1]$.*

**Proof.** From $T_1$ we have $T(0,1) = 0$. Then from $T_4$ it follows that

$$T(0,x) \leq T(0,1) = 0, \forall x \in [0,1],$$

i.e., $T(0,x) = 0$. Then from $T_2$ we get $T(x,0) = 0$.
    Similarly, we have $S(1,0) = 1$ so,

$$1 = S(1,0) \leq S(1,x),$$

and then $S(1,x) = 1 = S(x,1)$, $\forall x \in [0,1]$.    ∎

**Proposition 2.12.** *We have $T(x,y) \leq x \wedge y$ and $S(x,y) \geq x \vee y$ for any t-norm $T$, t-conorm $S$, and any $x,y \in [0,1]$.*

**Proof.** We have

$$T(x,y) \leq T(x,1) = x.$$

Also,

$$T(x,y) = T(y,x) \leq T(y,1) = y.$$

As a conclusion $T(x,y) \leq x \wedge y$. Similarly, $S(x,y) \geq x \vee y$.    ∎

The connection between t-norms and t-conorms is made by strong negations as follows.

**Definition 2.13.** *(Zadeh [154]) A triplet $(S,T,N)$ is called a **De Morgan triplet** if $T$ is a t-norm, $S$ is a t-conorm, $N$ is a strong negation, and if they fulfill De Morgan's law*

$$S(x,y) = N(T(N(x),N(y))).$$

**Remark 2.14.** *If $(S,T,N)$ is a De Morgan triplet then we have $T(x,y) = N(S(N(x),N(y)))$. Indeed,*

$$S(N(x),N(y)) = N(T(N(N(x)),N(N(y)))) = N(T(x,y)).$$

*Then*

$$N(S(N(x),N(y))) = N(N(T(x,y)) = T(x,y).$$

In what follows we give some examples of De Morgan triplets.

**Example 2.15.** *(Zadeh [154])The minimum and maximum*

$$x \wedge y = \min\{x,y\}$$
$$x \vee y = \max\{x,y\}$$

*together with the standard negation $N(x) = 1 - x$, form a De Morgan triplet. This De Morgan triplet (often called Gödel t-norm and t-conorm) plays a special role in fuzzy logic, and it was proposed to be used for fuzzy sets by Zadeh in [154]. Let us also observe that $\wedge$ is the greatest t-norm and $\vee$ is the least t-conorm as it was shown in Proposition 2.12*

**Example 2.16.** *(see e.g., Fodor-Roubens [60]) The product and the probabilistic sum*

$$x T_G y = x \cdot y$$
$$x S_G y = x + y - xy$$

*together with the standard negation $N(x) = 1 - x$, form a De Morgan triplet. These are called Goguen's t-norm and t-conorm.*

**Example 2.17.** *(Hajek [76])The Lukasiewicz t-norm and t-conorm*

$$xT_Ly = (x + y - 1) \vee 0$$
$$xS_Ly = (x + y) \wedge 1$$

*together with the standard negation $N(x) = 1 - x$, form a De Morgan triplet that plays a very special role in fuzzy logic, and the algebraic structure over the $[0, 1]$ interval generated by these operations is called an MV algebra (MV stands for Multivalued, see Cignoli-D'Ottaviano-Mundici, [38], Di Nola-Lettieri [47], Hajek [76]).*

**Example 2.18.** *(see e.g., Fodor [59]) The nilpotent minimum and nilpotent maximum with the standard negation is a De Morgan triplet:*

$$xT_0y = \begin{cases} x \wedge y & \text{if } x + y > 1 \\ 0 & \text{otherwise} \end{cases}$$
$$xS_0y = \begin{cases} x \vee y & \text{if } x + y < 1 \\ 1 & \text{otherwise} \end{cases} .$$

*These were introduced by Fodor and Perny and led to the study of similar fuzzy operations called uninorms and absorbing norms Rudas-Pap-Fodor [128].*

**Example 2.19.** *(see e.g. Klement-Mesiar-Pap [87])The drastic product and sum with the standard negation is a De Morgan triplet:*

$$xT_Dy = \begin{cases} x \wedge y & \text{if } x \vee y = 1 \\ 0 & \text{otherwise} \end{cases}$$
$$xS_Dy = \begin{cases} x \vee y & \text{if } x \wedge y = 0 \\ 1 & \text{otherwise} \end{cases} .$$

*This pair is also very important, because $T_D$ is the least t-norm and $S_D$ is the greatest t-conorm. Therefore, $T_D \leq T \leq \wedge$ and $\vee \leq S \leq S_D$ for any t-norm $T$ and t-conorm $S$.*

**Example 2.20.** *(see e.g. Klement-Mesiar-Pap [89]) There are several parametric families of t-norms and t-conorms. Perhaps the most famous one is the Frank family:*

$$xT_F^s y = \log_s \left(1 + \frac{(s^x - 1)(s^y - 1)}{s - 1}\right)$$
$$xS_F^s y = 1 - (1 - x)T_F^s(1 - y),$$

*where $s \in (0, \infty) \setminus \{1\}$. By definition Frank's t-norm and t-conorm form a De Morgan triplet with the standard negation. The importance of these lies in the fact that they satisfy the functional equation*

$$xT_F^s y + xS_F^s y = x + y.$$

## 2.3 Archimedean t-Norms and t-Conorms

Let $T$ be a t-norm and $S$ be a t-conorm. Given $x \in [0,1]$ we define

$$x_T^{(n)} = T(\underbrace{x, ..., x}_{n-times}) = T(x_T^{(n-1)}, x),\ n \geq 2$$

$$x_S^{(n)} = S(\underbrace{x, ..., x}_{n-times}) = S(x_S^{(n-1)}, x),\ n \geq 2.$$

**Definition 2.21.** *(see e.g. Klement-Mesiar-Pap [90], Fodor [59]) (i) A t-norm $T$ (and respectively a t-conorm $S$) is said to be continuous if it is continuous as a function on the unit interval.*

*(ii) A t-norm $T$ and respectively a t-conorm $S$ is said to be **Archimedean** if*

$$\lim_{n \to \infty} x_T^{(n)} = 0,$$

*and respectively*

$$\lim_{n \to \infty} x_S^{(n)} = 1,$$

*for any $x \in (0,1)$.*

**Proposition 2.22.** *(i) If $T$ is Archimedean then $T(x,x) < x, \forall x \in (0,1)$.*
*(ii) If $S$ is Archimedean then $S(x,x) > x, \forall x \in (0,1)$.*

**Proof.** (i) Suppose the contrary, i.e., that $x_T^{(2)} = T(x,x) \geq x$ for some $x \in (0,1)$. Then we have

$$x_T^{(3)} = T(T(x,x),x) \geq T(x,x) \geq x$$

and by induction we also have

$$x_T^{(n)} = T(\underbrace{x, ..., x}_{n-times}) \geq x.$$

Then $\lim_{n \to \infty} x_T^{(n)} \geq x$, which is a contradiction.
(ii) The proof of (ii) is similar. ∎

**Remark 2.23.** *Generally an Archimedean t-norm (or t-conorm) does not need to be continuous. If a t-norm (or t-conorm) is continuous and Archimedean then we obtain the following characterization:*

**Theorem 2.24.** *(i) Let $T$ be a continuous t-norm. Then $T$ is Archimedean if and only if*

$$T(x,x) < x, \forall x \in (0,1).$$

*(ii) Let $S$ be a continuous t-conorm. Then $S$ is Archimedean if and only if*

$$S(x,x) > x, \forall x \in (0,1).$$

**Proof.** (i) "⇒" If $T$ is Archimedean then the previous proposition ensures the property $T(x,x) < x, \forall x \in (0,1)$.

"⇐" Let us suppose now that $T$ fulfills the condition $T(x,x) < x, \forall x \in (0,1)$ and that $T$ is continuous. Then we have

$$x_T^{(n)} = T(\underbrace{x,...,x}_{n-times}) = T(T(x,x), \underbrace{x,...,x}_{n-2-times}) < T(\underbrace{x,...,x}_{n-1-times}) = x_T^{(n-1)},$$

so the sequence $x_T^{(n)}$ is decreasing (strictly). Also, $x_T^{(n)}$ is bounded from below by 0. So, $x_T^{(n)}$ is a convergent sequence. Let $y = \lim_{n\to\infty} x_T^{(n)}$. Let us suppose now that $y > 0$. Then

$$x_T^{(2n)} = T(x_T^{(n)}, x_T^{(n)})$$

and since $T$ is continuous we have

$$\lim_{n\to\infty} x_T^{(2n)} = \lim_{n\to\infty} T(x_T^{(n)}, x_T^{(n)}) = T(\lim_{n\to\infty} x_T^{(n)}, \lim_{n\to\infty} x_T^{(n)}).$$

Taking into account that $\lim_{n\to\infty} x_T^{(2n)}$ is also $y$, we get $y = T(y,y)$. But from the hypothesis, $T(y,y) < y$, so we get a contradiction.

(ii) The proof of (ii) is similar.    ∎

**Example 2.25.** *The product t-norm $T_G(x,y) = x \cdot y$ is Archimedean, while the min t-norm $T_M(x,y) = x \wedge y$ is not Archimedean.*

**Definition 2.26.** *(Klement-Mesiar-Pap [89]) Let $f : [a,b] \to [c,d]$ be monotonic, and $[a,b], [c,d] \subset [-\infty, \infty]$. The **pseudo-inverse** of $f$ is defined as $f^{(-1)} : [c,d] \to [a,b]$,*

$$f^{(-1)}(y) = \begin{cases} \sup(\{x|f(x) < y\} \cup \{a\}) & \text{if } f(a) < f(b) \\ \sup(\{x|f(x) > y\} \cup \{a\}) & \text{if } f(a) > f(b) \\ a & \text{if } f(a) = f(b) \end{cases}.$$

**Example 2.27.** *Let $f : [-1,1] \to [c,d]$, $[1.5, 2.5] \subset [c,d]$, and $f(x) = \frac{x+4}{2}$. Then we have*

$$f^{(-1)}(y) = \max\{\min(2y - 4, 1), -1\}.$$

*Indeed, first we observe that the image of the function $f$, denoted by $Im(f)$ is $Im(f) = f[-1,1] = [1.5, 2.5]$. Then*

$$f^{(-1)}(y) = \sup\left\{x \in [-1,1] | \frac{x+4}{2} < y\right\} \cup \{-1\}$$

$$= \sup\{x \in [-1,1] | x < 2y - 4\} \cup \{-1\}.$$

*If $2y - 4 < -1$ then $f^{(-1)}(y) = -1$. If $-1 \le 2y - 4 \le 1$ then $f^{(-1)}(y) = 2y - 4$. If $1 < 2y - 4$ then $f^{(-1)}(y) = 1$.*

*Also we observe that*

$$\max\{\min(2y - 4, 1), -1\} = \begin{cases} -1 & \text{if } 2y - 4 < -1 \\ 2y - 4 & \text{if } -1 \le 2y - 4 \le 1 \\ 1 & \text{if } 1 < 2y - 4 \end{cases}.$$

**Lemma 2.28.** *Let $f : [0,1] \to [0,\infty]$, $f(1) = 0$ be strictly decreasing and continuous. Then*

$$f^{(-1)}(y) = \begin{cases} f^{-1}(y) & \text{if } y \le f(0) \\ 0 & \text{otherwise} \end{cases},$$

*where $f^{-1}$ denotes the inverse function for the restriction $f : [0,1] \to Im(f)$.*

**Proof.** Indeed, $Im(f) = [0, f(0)]$, and $f$ is invertible on $Im(f)$ having its inverse $f^{-1}$. Then if $y \in Im(f)$,

$$f^{(-1)}(y) = \sup\{x | f(x) > y\} = \sup\{x | f(x) > f(f^{-1}(y))\}$$

$$= \sup\{x | x < f^{-1}(y)\} = f^{-1}(y).$$

If $y > f(0)$ then $f(x) > y > f(0) \iff x < 0$. Then $\sup\{x \in [0,1] | f(x) > y\} = \emptyset$ and then $\sup\{x | f(x) > y\} \cup \{0\} = 0$. ∎

**Lemma 2.29.** *If $f : [0,1] \to [0,\infty]$ is decreasing then*
*(i) $f^{(-1)}(f(x)) = x$*
*(ii) $f(f^{(-1)}(x)) = \min\{x, f(0)\}$.*

**Proof.** Since $f(x) \le f(0)$ we have $f^{(-1)}(f(x)) = f^{-1}(f(x)) = x$.
If $x \le f(0)$ then $f(f^{(-1)}(x)) = f(f^{-1}(x)) = x$. If $x > f(0)$ then $f^{(-1)}(x) = 0$,and so $f(f^{(-1)}(x)) = f(0)$ ∎

**Lemma 2.30.** *(see Aczél [1]) The only continuous solution $F : \mathbb{R} \to \mathbb{R}$ of Cauchy's functional equation*

$$F(x + y) = F(x) + F(y)$$

*is $F(x) = cx$, for some constant $c \in \mathbb{R}$.*

**Proof.** We have $F(0) = 2F(0)$, and so $F(0) = 0$. Then, $F(n) = nF(1)$. Let $F(1) = c$. It is easy to prove that $F(\frac{1}{n}) = \frac{F(1)}{n}$ and then $F(\frac{m}{n}) = \frac{m}{n}F(1)$, i.e., $F(r) = cr, \forall r \in \mathbb{Q}$. Then we extend $F$ by continuity and we get $F(x) = cx$, for any $x \in \mathbb{R}$. ∎

**Theorem 2.31.** *(Representation of Archimedean t-norms, see e.g. Klement-Mesiar-Pap, [87], Klement-Mesiar-Pap, [88]) For a continuous function $T : [0,1]^2 \to [0,1]$ the following statements are equivalent:*

*(i) $T$ is a continuous Archimedean t-norm*
*(ii) $T$ has a continuous additive generator, i.e., there is a continuous strictly decreasing $t : [0,1] \to [0,\infty]$, $t(1) = 0$, which is uniquely determined up to a multiplicative constant such that for all $x, y \in [0,1]$ we have*

$$T(x,y) = t^{(-1)}(t(x) + t(y)).$$

**Proof.** (ii)$\Rightarrow$(i). Let $T(x,y) = t^{(-1)}(t(x) + t(y))$. We have to prove that it is a continuous Archimedean t-norm.

T1: From the previous lemma we get $T(x,1) = t^{(-1)}(t(x) + t(1)) = t^{(-1)}(t(x)) = x$.

T2: $T(x,y) = T(y,x)$ is obvious.

T3: We have

$$T(T(x,y),z) = t^{(-1)}(t^{(-1)}(t(x) + t(y)) + t(z))$$

$$= t^{(-1)}(\min(t(x) + t(y), t(0)) + t(z))$$

$$= \min(\min(t(x) + t(y), t(0)) + t(z), t(0)).$$

If $t(x) + t(y) > t(0)$ then

$$T(T(x,y),z) = \min(t(0) + t(z), t(0)) = t(0)$$

$$= \min(t(x) + t(y) + t(z), t(0)) = T(x,T(y,z)).$$

If $t(x) + t(y) \le t(0)$ then

$$T(T(x,y),z) = \min(t(x) + t(y) + t(z), t(0)) = T(x,T(y,z)).$$

T4: To prove that $T$ is increasing we consider $x \le u$ and $y \le v$. Then $t(x) \ge t(u)$ and $t(y) \ge t(v)$ and $t(x) + t(y) \ge t(u) + t(v)$. Since $t^{(-1)}$ is decreasing we obtain

$$T(x,y) = t^{(-1)}(t(x) + t(y)) \le t^{(-1)}(t(u) + t(v)) = T(u,v).$$

$T$- continuous: Since $t$ is continuous and since $t^{-1}$ is continuous we get $t^{(-1)}$ continuous when $x \ne f(0)$. Also,

$$\lim_{x \nearrow t(0)} t^{(-1)}(x) = \lim_{x \nearrow t(0)} t^{-1}(x) = t^{-1}(t(0)) = 0.$$

$T$- Archimedean:

$$T(x,x) = t^{(-1)}(2t(x)) < t^{(-1)}(t(x)) = \min\{x, t(0)\} \le x,$$

i.e., $T(x,x) < x, \forall x \in (0,1)$.

(i)$\Rightarrow$(ii) Now let $T$ be a continuous Archimedean t-norm. As before let $x_T^{(n)} = T(\underbrace{x,...,x}_{n-times}) = T(x, x_T^{(n-1)})$, $x_T^{(0)} = 1$. Then we define

$$x_T^{(\frac{1}{n})} = \sup\{y \in [0,1] | y_T^{(n)} < x\}$$

and

$$x_T^{(\frac{m}{n})} = \left(x_T^{(\frac{1}{n})}\right)_T^{(m)}.$$

Since $T$ is Archimedean we have $\lim_{n \to \infty} x_T^{(\frac{1}{n})} = 1$ and $x_T^{(\frac{m}{n})}$ is well defined since $x_T^{(\frac{m}{n})} = x_T^{(\frac{km}{kn})}$.

Let $a$ be an arbitrary positive number. Let $h : \mathbb{Q} \cap [0, \infty] \to [0, 1]$, $h(r) = a^{(r)}$. Since $T$ is continuous then $h$ is continuous. Also, it is easy to check that

$$x_T^{(\frac{m}{n} + \frac{p}{q})} = T(x_T^{(\frac{m}{n})}, x_T^{(\frac{p}{q})}),$$

so we have

$$h(r + s) = a_T^{(r+s)} = T(a_T^{(r)}, a_T^{(s)}) \le a_T^{(r)} = h(r),$$

so $h$ is non-increasing. Moreover, $h$ is strictly decreasing

$$h\left(\frac{m}{n} + \frac{p}{q}\right) = h\left(\frac{mq + np}{nq}\right) \le h\left(\frac{mq + 1}{nq}\right)$$

$$= \left(a_T^{(\frac{1}{nq})}\right)_T^{(mq+1)} < \left(a_T^{(\frac{1}{nq})}\right)_T^{(mq)} = h\left(\frac{m}{n}\right).$$

Since $h$ is monotone and continuous we can uniquely extend it to a real function $\bar{h} : [0, \infty] \to [0, 1]$

$$\bar{h}(x) = \inf\{h(r) : r \in \mathbb{Q} \cap [0, x]\}.$$

Since $T$ is continuous $\bar{h}(x + y) = T(\bar{h}(x), \bar{h}(y))$. We define

$$t(x) = \sup\{y \in [0, \infty] | \bar{h}(y) > x\} \cup \{0\}.$$

We observe that $t$ is the pseudo-inverse of $\bar{h}$.

Then we obtain

$$T(x, y) = T(\bar{h}(t(x)), \bar{h}(t(y))) = \bar{h}(t(x) + t(y))$$

$$= t^{(-1)}(t(x) + t(y)),$$

i.e., we have obtained the existence of the additive generator $t$.

To prove the uniqueness of $t$ let us suppose that there are two additive generators $t_1, t_2$. Then we have

$$t_1^{(-1)}(t_1(x) + t_1(y)) = t_2^{(-1)}(t_2(x) + t_2(y)).$$

Let $u = t_2(x)$, $v = t_2(y)$. Then

$$t_1^{(-1)}(t_1 \circ t_2^{(-1)}(u) + t_1 \circ t_2^{(-1)}(v)) = t_2^{(-1)}(u + v).$$

If $u, v \in [0, t_2(0)]$ and $u + v \in [0, t_2(0)]$ then

$$t_1 \circ t_2^{(-1)}(u) + t_1 \circ t_2^{(-1)}(v) = t_1 \circ t_2^{(-1)}(u+v),$$

i.e., $t_1 \circ t_2^{(-1)}$ satisfies Cauchy's functional equation $F(u+v) = F(u) + F(v)$ which has its solution $F(x) = bx$ with $b$ being a constant in $(0, \infty)$. Then $t_1 \circ t_2^{(-1)}(x) = bx, x \in [0, t_2(0)]$. Recall that $x = t_2(y)$ and we get $t_1(y) = bt_2(y), \forall y \in [0,1]$, so $t$ is unique up to a multiplicative constant and the proof is complete.   ■

**Theorem 2.32.** *(Representation of Archimedean t-conorms, e.g., Klement-Mesiar-Pap [87]) For a continuous function* $S : [0,1]^2 \to [0,1]$ *the following affirmations are equivalent:*

*(i)* $S$ *is a continuous Archimedean t-conorm*

*(ii)* $S$ *has a continuous additive generator, i.e., there is a continuous strictly increasing* $s : [0,1] \to [0, \infty]$, $s(0) = 0$, *which is uniquely determined up to a multiplicative constant such that for all* $x, y \in [0,1]$ *we have*

$$S(x,y) = s^{(-1)}(s(x) + s(y)).$$

We can write $T$ and $S$ as

$$T(x,y) = t^{-1}(\min\{t(x) + t(y), t(0)\})$$

$$S(x,y) = s^{-1}(\max\{s(x) + s(y), s(0)\})$$

**Example 2.33.** *By considering* $t(x) = -\ln x$ *as additive generator we obtain the t-norm* $T(x,y) = x \cdot y$ *(product t-norm).*

*By taking* $t(x) = 1 - x$ *we obtain the Lukasiewicz t-norm* $T(x,y) = \max\{x + y - 1, 0\}$.

*For Frank's t-norm*

$$x T_F^s y = \log_s \left( 1 + \frac{(s^x - 1)(s^y - 1)}{s - 1} \right)$$

*the additive generator is*

$$t_s(x) = \begin{cases} -\ln x & \text{if} \quad s = 1 \\ -\ln \frac{s^x - 1}{s - 1} & s \in (0, \infty) \setminus \{1\} \end{cases}.$$

## 2.4   Fuzzy Implications

Fuzzy implications were studied by several authors and their usefulness in fuzzy inference systems makes them very important.

**Definition 2.34.** *(Fodor [58]) Let $I : [0,1] \times [0,1] \to [0,1]$ be a function. If the following conditions are fulfilled*

*I1: If $x \leq y$, then $I(x,z) \geq I(y,z)$, i.e., $I$ is decreasing in its first variable;*

*I2: If $y \leq z$, then $I(x,y) \leq I(x,z)$, i.e., $I$ is increasing in its second variable;*

*I3: $I(1,0) = 0$, $I(0,0) = I(1,1) = 1$*

*then $I$ is called a **fuzzy implication**.*

**Example 2.35.** *The operations $I_1(x,y) = \max\{1-x,y\}$*
*$I_2(x,y) = \min\{1-x+y,1\}$ and*
$$I_3(x,y) = \begin{cases} 1 & \text{if } x \leq y \\ \frac{y}{x} & \text{if } x > y \end{cases} \quad \text{are fuzzy implications.}$$

A fuzzy implication is often denoted by $x \to y$.

**Proposition 2.36.** *If $I$ is a fuzzy implication and if $N$ is a negation then $I'(x,y) = I(N(y),N(x))$ is a fuzzy implication too.*

**Proof.** The proof is left as an exercise. ∎

**Proposition 2.37.** *If $I$ is a fuzzy implication then*
*(i) $I(0,x) = 1, \forall x \in [0,1]$;*
*(ii) $I(x,1) = 1, \forall x \in [0,1]$.*

**Proof.** (i):
$$1 = I(0,0) \leq I(0,x) \leq I(0,1).$$

Then it follows that $I(0,x) = 1$ for any $x \in [0,1]$, and also $I(0,1) = 1$.
(ii):
$$1 = I(0,1) \geq I(x,1) \geq I(1,1) = 1.$$

Then $I(x,1) = 1, \forall x \in [0,1]$. ∎

**Definition 2.38.** *Let $S$ be a t-conorm and $N$ be a strong negation. Then*

$$I(x,y) = S(N(x),y)$$

*is called an S- **implication**.*

**Proposition 2.39.** $I(x,y) = S(N(x),y)$ *is an implication.*

**Proof.** Let us suppose that $x \leq z$. Then $N(x) \geq N(z)$ and then $S(N(x),y) \geq S(N(z),y)$, so $I$ is decreasing in the first argument. The fact that $I$ is increasing in the second argument is obvious. We observe that

$$I(1,0) = S(N(1),0) = S(0,0) = 0,$$

$I(0,0) = S(1,0) = 1$ and $I(1,1) = S(0,1) = 1$. ∎

**Example 2.40.** *(Kleene-Dienes) Let $S = \max$, $N(x) = 1-x$. Then the associated S-implication is $I(x,y) = \max\{1-x,y\}$.*

**Example 2.41.** *(Reichenbach) Let $S$ be the probabilistic sum, i.e. $S(x,y) = x + y - xy$, and $N(x) = 1 - x$. Then $I(x,y) = 1 - x + xy$ is the S-implication associated to $S$.*

**Example 2.42.** *(Lukasiewicz) Let $S(x,y) = \min\{x+y,1\}$ and consider the standard negation. Then we obtain the S-implication $I(x,y) = \min\{1 - x + y, 1\}$.*

**Remark 2.43.** *If $I(x,y)$ is an S-implication then $I(N(y), N(x))$ is also an S-implication.*

**Definition 2.44.** *Let $T$ be a t-norm.*

$$I_T(x,y) = \sup\{z | T(x,z) \leq y\}$$

*is called a **residual implication** (an **R-implication**).*

**Proposition 2.45.** *An residual implication $I_T$ is an implication.*

**Proof.** Let $x_1 \leq x_2$. Then $T(x_1, z) \leq T(x_2, z), \forall z \in [0,1]$. If $z_0 \in \{z | T(x_2, z) \leq y\}$ then $z_0 \in \{z | T(x_1, z) \leq y\}$, i.e.,

$$\{z | T(x_2, z) \leq y\} \subseteq \{z | T(x_1, z) \leq y\}.$$

Then

$$I_T(x_2, y) = \sup\{z | T(x_2, z) \leq y\}$$

$$\leq \sup\{z | T(x_1, z) \leq y\} = I_T(x_1, y).$$

The verification of the other properties of an implication is straight-forward. ∎

**Remark 2.46.** *We have $I_T(x,x) = 1$, $\forall x \in [0,1]$. Indeed, since $I_T(x,x) = \sup\{z | T(x,z) \leq x\}$ and since $T(x,1) = x$ we have $I_T(x,x) = 1$.*

**Example 2.47.** *$T = \min$ gives the Gödel implication*

$$I_T(x,y) = \begin{cases} 1 & if \quad x \leq y \\ y & if \quad x > y \end{cases}.$$

*$T = product$ gives the Goguen implication*

$$I_T(x,y) = \begin{cases} 1 & if \quad x \leq y \\ \frac{y}{x} & if \quad x > y \end{cases}.$$

*$T = Lukasiewicz$ t-norm, gives the R-implication $I_T(x,y) = \min\{1 - x + y, 1\}$.*

## 2.5   Fuzzy Equivalence

**Definition 2.48.** *(Fodor-Yager [61]) A function $E : [0,1]^2 \to [0,1]$ is a* **fuzzy equivalence** *if it satisfies the conditions:*
  E1: $E(x, y) = E(y, x)$.
  E2: $E(0, 1) = E(1, 0) = 0$.
  E3: $E(x, x) = 1, \forall x \in [0, 1]$.
  E4: *If* $x \leq x' \leq y' \leq y$ *then* $E(x, y) \leq E(x', y')$.

A fuzzy equivalence is often denoted by $x \leftrightarrow y$.

**Theorem 2.49.** *(Fodor-Yager [61]) The following statements are equivalent:*
  *(a) $E$ is a fuzzy equivalence.*
  *(b) There exists a fuzzy implication $I$ with the property $I(x, x) = 1, \forall x \in [0, 1]$ such that*

$$E(x, y) = \min\{I(x, y), I(y, x)\}.$$

  *(c) There exists a fuzzy implication $I$ with the property $I(x, x) = 1, \forall x \in [0, 1]$ such that*

$$E(x, y) = I(\max\{x, y\}, \min\{x, y\}).$$

**Proof.** (b)⇒(a). Let $E(x, y) = \min\{I(x, y), I(y, x)\}$. Then we can easily check the properties E1-E3. Regarding E4 we observe that if $x \leq x' \leq y' \leq y$ then we have

$$I(y, x) \leq I(x, x) \leq I(x, y)$$

and so $E(x, y) = I(y, x)$. Similarly $E(x', y') = I(y', x')$. Also, $I(y', x') \geq I(y', x) \geq I(y, x)$. Then $E(x, y) \leq E(x', y')$.
  (a)⇒(b). We consider

$$I(x, y) = \left\{ \begin{array}{ll} 1 & \text{if } x \leq y \\ E(x, y) & \text{if } x > y \end{array} \right. .$$

Then $I$ is a fuzzy implication. We show that $I$ is decreasing in the first argument. Let $x \leq z$. If $x \leq y$ then $I(x, y) = 1 \geq I(z, y)$. If $x > y$ then $y < x \leq z$ and then

$$I(x, y) = E(x, y) \geq E(z, y) = I(z, y).$$

Now we prove that $I$ is increasing in the second argument. Let $y \leq z$. If $x > z$ then

$$I(x, y) = E(x, y) \leq E(x, z) = I(x, z).$$

If $x \leq z$ then $I(x, z) = 1$ so $I(x, y) \leq I(x, z)$. The boundary conditions for the fuzzy implications can easily be verified.
  Now we prove that

$$E(x, y) = \min\{I(x, y), I(y, x)\}.$$

Indeed, since

$$I(y,x) = \begin{cases} 1 & \text{if } y \le x \\ E(y,x) = E(x,y) & \text{if } y > x \end{cases}$$

we observe that

$$E(x,y) = \begin{cases} I(x,y) & \text{if } x > y \\ I(y,x) & \text{if } x \le y \end{cases}.$$

Also, we observe that if $x > y$ then

$$\min\{I(x,y), I(y,x)\} = I(x,y).$$

Indeed, if $x > y$ then

$$I(x,y) \le I(y,y) \le I(y,x).$$

The case $x < y$ can be discussed in a similar way. If $x = y$ then $I(x,y) = I(y,x) = 1$.

The equivalence (b) $\Longleftrightarrow$ (c) is obvious based on the fact that if $x > y$ then $\max\{x,y\} = x$ and $\min\{x,y\} = y$. Also, in this case

$$I(x,y) \le I(y,y) \le I(y,x)$$

and

$$E(x,y) = I(x,y) = I(\max\{x,y\}, \min\{x,y\}).$$

When $x < y$ a similar reasoning applies. If $x = y$ then

$$E(x,y) = I(x,y) = I(\max\{x,y\}, \min\{x,y\}) = 1.$$

$\blacksquare$

**Example 2.50.** *The fuzzy equivalence $x \leftrightarrow y = \min\{1 - x + y, 1 - y + x\}$ is derived from the Lukasiewicz t-norm.*

## 2.6  Problems

1. Verify that the $\lambda$-complement $N_\lambda(x) = \frac{1-x}{1+\lambda x}$, $\lambda > -1$, is a strong negation. Find a function $\varphi : [0,1] \to [0,1]$. Such that $N_\lambda(x) = \varphi^{-1}(1 - \varphi(x))$.

2. Prove that the product and the probabilistic sum:

$$x T_G y = x \cdot y$$
$$x S_G y = x + y - xy.$$

is a t-norm and a t-conorm respectively. Decide whether they form a DeMorgan triplet with the standard negation.

3. Prove that the Lukasiewicz operations

$$xT_Ly = (x + y - 1) \vee 0$$
$$xS_Ly = (x + y) \wedge 1.$$

are a t-norm and a t-conorm respectively. Decide whether they form a DeMorgan triplet together with the standard negation.

4. Show that the nilpotent minimum and nilpotent maximum are t-norm and t-conorm respectively

$$xT_0y = \begin{cases} x \wedge y & \text{if } x + y > 1 \\ 0 & \text{otherwise} \end{cases}$$

$$xS_0y = \begin{cases} x \vee y & \text{if } x + y < 1 \\ 1 & \text{otherwise} \end{cases}.$$

Decide whether they form a De Morgan triplet with the standard negation.

5. Consider the fuzzy sets $A, B : \{1, 2, ..., 10\} \rightarrow [0, 1]$ defined as $A(x) = \frac{0}{1} + \frac{0.2}{2} + \frac{0.7}{3} + \frac{1}{4} + \frac{0.7}{5} + \frac{0.2}{6} + \frac{0}{7} + \frac{0}{8} + \frac{0}{9} + \frac{0}{10}$, $B(x) = \frac{0}{1} + \frac{0}{2} + \frac{0}{3} + \frac{0.3}{4} + \frac{0.5}{5} + \frac{0.8}{6} + \frac{1}{7} + \frac{0.5}{8} + \frac{0.2}{9} + \frac{0}{10}$,. Calculate $AT_LB$, $AS_LB$, $AT_GB$, $AS_GB$, where $T_{L,G}$ and $S_{L,G}$ stand for the Lukasiewicz and Goguen t-norms and t-conorms.

6. Consider the fuzzy sets $A, B : \mathbb{R}_+ \rightarrow [0, 1]$ defined as $A(x) = \frac{1}{1+x^2}$, $B(x) = \frac{1}{10^x}$. Calculate $AT_LB$, $AS_LB$, $AT_GB$, $AS_GB$, where $T_{L,G}$ and $S_{L,G}$ stand for the Lukasiewicz and Goguen t-norms and t-conorms, and graph them.

7. Let us consider the fuzzy sets

$$A(x) = \begin{cases} 0 & \text{if,} & x < 2 \\ \frac{x-2}{4} & \text{if} & 2 \leq x < 6 \\ \frac{10-x}{4} & \text{if} & 6 \leq x < 10 \\ 0 & \text{if} & 10 \leq x \end{cases},$$

$$B(x) = \begin{cases} 0 & \text{if,} & x < 3 \\ x - 3 & \text{if} & 3 \leq x < 4 \\ 1 & \text{if} & 4 \leq x < 5 \\ \frac{7-x}{2} & \text{if} & 5 \leq x \leq 7 \\ 0 & \text{if} & 7 < x \end{cases}.$$

Calculate $AT_LB$, $AS_LB$, $AT_GB$, $AS_GB$, where $T_{L,G}$ and $S_{L,G}$ stand for the Lukasiewicz and Goguen t-norms and t-conorms, and graph them.

8. Prove that Frank's t-norm and t-conorm

$$xT_F^s y = \log_s \left(1 + \frac{(s^x - 1)(s^y - 1)}{s - 1}\right)$$
$$xS_F^s y = 1 - (1 - x)T_F^s(1 - y),$$

where $s \in (0, \infty) \setminus \{1\}$ verify the relation

$$xT_F^s y + xS_F^s y = x + y.$$

9. Prove that the drastic product and sum are the least t-norm and the greatest t-conorm respectively.

10. Prove that any t-norm $T$ and any t-conorm $S$ is distributive with respect to $\vee$ and $\wedge$ i.e.,

$$(x \vee y)Tz = (xTz) \vee (yTz)$$

$$(x \wedge y)Tz = (xTz) \wedge (yTz)$$

$$(x \vee y)Sz = (xSz) \vee (ySz)$$

$$(x \wedge y)Sz = (xSz) \wedge (ySz)$$

for any $x, y, z \in [0, 1]$.

11. Let us consider the fuzzy sets

$$A(x) = \begin{cases} 0 & \text{if,} & x < 2 \\ \frac{x-2}{5} & \text{if} & 2 \leq x < 7 \\ \frac{10-x}{3} & \text{if} & 7 \leq x < 10 \\ 0 & \text{if} & 10 \leq x \end{cases},$$

$$B(x) = \begin{cases} 0 & \text{if,} & x < 3 \\ x - 3 & \text{if} & 3 \leq x < 4 \\ 1 & \text{if} & 4 \leq x < 5 \\ \frac{7-x}{2} & \text{if} & 5 \leq x \leq 7 \\ 0 & \text{if} & 7 < x \end{cases}.$$

Consider the fuzzy implications

$$I_1(x, y) = \max\{1 - x, y\}$$

and

$$I_2(x, y) = \begin{cases} 1 & \text{if} & x \leq y \\ y & \text{if} & x > y \end{cases}.$$

Calculate and graph $I_1(A, B)$ and $I_2(A, B)$.

12. Consider the fuzzy equivalences

$$E_1 = \min\{I_1(x, y), I_1(y, x)\}$$

and

$$E_2 = \min\{I_2(x, y), I_2(y, x)\}$$

generated by the implications in the previous problem. Calculate and graph $E_1(A, B)$ and $E_2(A, B)$ based on the fuzzy sets considered in Problem 11.

13. Prove that the implications in Example 2.35 are fuzzy implications.

14. Prove Proposition 2.36.

15. Prove that the function in Example 2.50 is a fuzzy equivalence.

16. Find the fuzzy R-implication, S-implication and the associated fuzzy equivalences that we can obtain starting from Goguen and Fodor-Perny t-norm and t-conorm respectively.

# 3
# Fuzzy Relations

## 3.1 Fuzzy Relations

**Definition 3.1.** *(Classical relation). A subset $R \subseteq X \times Y$ where $X$ and $Y$ are classical sets, is a classical relation.*

A classical relation can be characterized by a function $R : X \times Y \to \{0, 1\}$,

$$R(x, y) = \begin{cases} 1 & \text{if} \quad (x, y) \in R \\ 0 & \text{otherwise} \end{cases}.$$

**Definition 3.2.** *(Fuzzy Relation, Sanchez [129], Di Nola-Sessa-Pedrycz-Sanchez [48], De Baets [42]) Let $X, Y$ be two classical sets. A mapping $R : X \times Y \to [0, 1]$ is called a **fuzzy relation**. The number $R(x, y) \in [0, 1]$ can be interpreted as the degree of relationship between $x$ and $y$.*

**Remark 3.3.** *A fuzzy relation can be seen as a fuzzy subset of the set $X \times Y$. We denote by $\mathcal{F}(X \times Y)$ the family of all fuzzy relations between elements of $X$ and $Y$.*

**Example 3.4.** $R =$ *"much greater than"*

$$R(x, y) = \begin{cases} \frac{1}{1 + \frac{100}{(x - y)^2}} & \text{if} \quad x > y \\ 0 & \text{otherwise} \end{cases}.$$

B. Bede: *Mathematics of Fuzzy Sets and Fuzzy Logic*, STUDFUZZ 295, pp. 33–49.
DOI: 10.1007/978-3-642-35221-8_3    © Springer-Verlag Berlin Heidelberg 2013

A fuzzy relation between elements in two finite sets $X = \{x_1, x_2, ..., x_m\}$ and $Y = \{y_1, y_2, ..., y_n\}$ can be represented as a matrix

$$R = \begin{pmatrix} R(x_1, y_1) & R(x_1, y_2) & ... & R(x_1, y_n) \\ R(x_2, y_1) & R(x_2, y_2) & ... & R(x_2, y_n) \\ ... & ... & ... & ... \\ R(x_m, y_1) & R(x_m, y_2) & ... & R(x_m, y_n) \end{pmatrix}.$$

Since fuzzy relations are themselves fuzzy sets, it is possible to perform fuzzy set operations on them

$$N(R(x, y)) = \bar{R}(x, y) = 1 - R(x, y),$$

$$(R \vee S)(x, y) = R(x, y) \vee S(x, y),$$

$$(R \wedge S)(x, y) = R(x, y) \wedge S(x, y).$$

The inverse in a set-theoretic approach for a fuzzy relation would be

$$R^{-1}(x, y) = R(y, x),$$

i.e. the transpose. We can also consider t-norm and t-conorm operations

$$T(R, P)(x, y) = T(R(x, y), P(x, y)),$$

$$S(R, P)(x, y) = S(R(x, y), P(x, y)),$$

where $T, S$ are a t-norm and a t-conorm respectively.

## 3.2  Max-Min Composition

**Definition 3.5.** *Let $R \in \mathcal{F}(X \times Y)$ and $S \in \mathcal{F}(Y \times Z)$ be fuzzy relations. Then $R \circ S \in \mathcal{F}(X \times Z)$, defined as*

$$R \circ S(x, z) = \bigvee_{y \in Y} R(x, y) \wedge S(y, z),$$

*is called the **max-min composition** of the fuzzy relations $R$ and $S$.*

**Remark 3.6.** *The max-min composition is well defined being the supremum of a nonempty bounded subset of real numbers.*

If $R \in \mathcal{F}(X \times X)$ then we can define $R^2 = R \circ R$, and generally $R^n = R \circ R^{n-1}$, $n \geq 2$.

Let $X = \{x_1, ..., x_n\}$, $Y = \{y_1, ..., y_m\}$, and $Z = \{z_1, ..., z_p\}$ be finite sets. If $R = (r_{ij})_{i=1,...,n, j=1,...,m} \in \mathcal{F}(X \times Y)$, and $S = (s_{jk})_{j=1,...,m, k=1,...,p} \in \mathcal{F}(Y \times Z)$ are discrete fuzzy relations then the composition $T = (t_{ik})_{i=1,...,n, k=1,...,p} = R \circ S \in \mathcal{F}(X \times Z)$ is given by

$$t_{ik} = \bigvee_{j=1}^{m} r_{ij} \wedge s_{jk},$$

$i = 1, ..., n$, $k = 1, ..., p$.

**Example 3.7.** If $R = \begin{pmatrix} 0.3 & 0.7 & 0.2 \\ 1 & 0 & 0.9 \end{pmatrix}$ and $S = \begin{pmatrix} 0.8 & 0.3 \\ 0.1 & 0 \\ 0.5 & 0.6 \end{pmatrix}$ then

$R \circ S = \begin{pmatrix} 0.3 & 0.3 \\ 0.8 & 0.6 \end{pmatrix}$.

**Proposition 3.8.** *(i) The max-min composition is associative, i.e.,*

$$(R \circ S) \circ Q = R \circ (S \circ Q),$$

*where $R \in \mathcal{F}(X \times Y)$, $S \in \mathcal{F}(Y \times Z)$ and $Q \in \mathcal{F}(Z \times U)$.*
*(ii) Let $R_1, R_2 \in \mathcal{F}(X \times Y)$ and $Q \in \mathcal{F}(Y \times Z)$. If $R_1 \leq R_2$ then*

$$R_1 \circ Q \leq R_2 \circ Q.$$

**Proof.** The proof of (i) is left to the reader.
(ii)

$$R_1 \circ Q(x, z) = \bigvee_{y \in Y} R_1(x, y) \wedge Q(y, z)$$

$$\leq \bigvee_{y \in Y} R_2(x, y) \wedge Q(y, z) = R_2 \circ Q(x, z).$$

∎

**Proposition 3.9.** *For any $R, S \in \mathcal{F}(X \times Y)$ and $Q \in \mathcal{F}(Y \times Z)$ we have*
*(i) $(R \vee S) \circ Q = (R \circ Q) \vee (S \circ Q)$*
*(ii) $(R \wedge S) \circ Q \leq (R \circ Q) \wedge (S \circ Q)$.*

**Proof.** (i) We prove the equality by double inclusion.

$$(R \vee S) \circ Q(x, z) = \bigvee_{y \in Y} (R \vee S)(x, y) \wedge Q(y, z)$$

$$= \bigvee_{y \in Y} (R(x, y) \wedge Q(y, z)) \vee (S(x, y) \wedge Q(y, z))$$

$$\leq \bigvee_{y \in Y} (R(x, y) \wedge Q(y, z)) \vee \bigvee_{y \in Y} (S(x, y) \wedge Q(y, z)),$$

i.e.,
$$(R \vee S) \circ Q \leq (R \circ Q) \vee (S \circ Q).$$

On the other hand, from the previous proposition we have $R \circ Q \leq (R \vee S) \circ Q$ and $S \circ Q \leq (R \vee S) \circ Q$, and then

$$(R \circ Q) \vee (S \circ Q) \leq (R \vee S) \circ Q.$$

Combining the two inequalities, the required conclusion follows.

(ii) We have

$$(R \wedge S) \circ Q(x, z) = \bigvee_{y \in Y} (R \wedge S)(x, y) \wedge Q(y, z)$$

$$= \bigvee_{y \in Y} (R(x, y) \wedge Q(y, z)) \wedge (S(x, y) \wedge Q(y, z))$$

$$\leq \bigvee_{y \in Y} (R(x, y) \wedge Q(y, z)) \wedge \bigvee_{y \in Y} (S(x, y) \wedge Q(y, z))$$

$$= (R \circ Q) \wedge (S \circ Q)(x, z).$$

$\blacksquare$

**Remark 3.10.** *Equality in (ii) does not hold. Indeed, if we consider* $R = \begin{pmatrix} 1 & 0 \\ 1 & 1 \end{pmatrix}$, $S = \begin{pmatrix} 0 & 1 \\ 1 & 1 \end{pmatrix}$, $Q = \begin{pmatrix} 1 & 1 \\ 1 & 1 \end{pmatrix}$ *then* $(R \wedge S) \circ Q = \begin{pmatrix} 0 & 0 \\ 1 & 1 \end{pmatrix}$ *while* $(R \circ Q) \wedge (S \circ Q) = \begin{pmatrix} 1 & 1 \\ 1 & 1 \end{pmatrix}$.

## 3.3   Min-Max Composition

**Definition 3.11.** *(e.g. Nobuhara-Bede-Hirota [118]) Let* $R \in \mathcal{F}(X \times Y)$ *and* $S \in \mathcal{F}(Y \times Z)$. *Then* $R \bullet S \in \mathcal{F}(X \times Z)$, *defined as*

$$R \bullet S(x, z) = \bigwedge_{y \in Y} R(x, y) \vee S(y, z)$$

*is called the* **min-max composition** *of the fuzzy relations $R$ and $S$.*

**Example 3.12.** *If* $R = \begin{pmatrix} 0.3 & 0.7 & 0.2 \\ 1 & 0 & 0.9 \end{pmatrix}$ *and* $S = \begin{pmatrix} 0.8 & 0.3 \\ 0.1 & 0 \\ 0.5 & 0.6 \end{pmatrix}$ *then*

$$R \bullet S = \begin{pmatrix} 0.5 & 0.3 \\ 0.1 & 0 \end{pmatrix}.$$

**Proposition 3.13.** *(i) The min-max composition is associative, i.e., for any* $R \in \mathcal{F}(X \times Y)$, $S \in \mathcal{F}(Y \times Z)$ *and* $T \in \mathcal{F}(Z \times U)$ *we have*

$$(R \bullet S) \bullet T = R \bullet (S \bullet T).$$

*(ii) Consider* $R_1, R_2 \in \mathcal{F}(X \times Y)$, $Q \in \mathcal{F}(Y \times Z)$. *If* $R_1 \leq R_2$ *then*

$$R_1 \bullet Q \leq R_2 \bullet Q.$$

**Proof.** The proof is left to the reader as an exercise.  ∎

**Proposition 3.14.** *For any* $R, S \in \mathcal{F}(X \times Y)$ *and* $T \in \mathcal{F}(Y \times Z)$ *we have*
*(i)* $(R \wedge S) \bullet T = (R \bullet T) \wedge (S \bullet T)$.
*(ii)* $(R \vee S) \bullet T \geq (R \bullet T) \vee (S \bullet T)$.

**Proof.** The proof of the proposition is similar to the corresponding result in the max-min composition case and is left as an exercise.  ∎

**Remark 3.15.** *Equality in (ii) does not hold. Indeed,* $R = \begin{pmatrix} 1 & 0 \\ 1 & 1 \end{pmatrix}$, $S = \begin{pmatrix} 0 & 1 \\ 1 & 1 \end{pmatrix}$, $T = \begin{pmatrix} 0 & 0 \\ 0 & 0 \end{pmatrix}$ *then* $(R \vee S) \bullet T = \begin{pmatrix} 1 & 1 \\ 1 & 1 \end{pmatrix}$ *while* $(R \bullet T) \vee (S \bullet T) = \begin{pmatrix} 0 & 0 \\ 1 & 1 \end{pmatrix}$.

**Proposition 3.16.** *If we consider the standard negation we have* $\overline{R \circ S} = \bar{R} \bullet \bar{S}$ *and* $\overline{R \bullet S} = \bar{R} \circ \bar{S}$.

**Proof.** We observe that

$$\overline{R \circ S}(x, z) = \overline{\bigvee_{y \in Y} R(x, y) \wedge S(y, z)} = \bigwedge_{y \in Y} \overline{R(x, y) \wedge S(y, z)}$$

$$= \bigwedge_{y \in Y} \bar{R}(x, y) \vee \bar{S}(y, z) = \bar{R} \bullet \bar{S}(x, z).$$

The equality $\overline{R \bullet S} = \bar{R} \circ \bar{S}$. can be obtained in a similar way.  ∎

**Remark 3.17.** *The min-max composition can be naturally generalized to min-t-conorm compositions*

$$R \bullet_S P(x, z) = \bigwedge_{y \in Y} R(x, y) S P(y, z),$$

*where* $S$ *is an arbitrary t-conorm.*

## 3.4   Min → Composition

Let → be the standard **Gödel implication** defined as

$$x \to y = \sup\{z \in [0,1] | x \wedge z \le y\} = \begin{cases} 1 & \text{if} \quad x \le y \\ y & \text{otherwise} \end{cases}.$$

**Proposition 3.18.** *For any $x, y, z \in [0,1]$ we have*
*(i) $(x \vee y) \to z = (x \to z) \wedge (y \to z)$.*
*(ii) $(x \wedge y) \to z = (x \to z) \vee (y \to z)$.*
*(iii) $x \to (y \vee z) = (x \to y) \vee (x \to z)$.*
*(iv) $x \to (y \wedge z) = (x \to y) \wedge (x \to z)$.*
*(v) $x \wedge (x \to y) \le y$.*
*(vi) $x \to (x \wedge y) \ge y$.*
*(vii) $(x \to y) \to y \ge x$.*

**Proof.** (i) Case 1. If $x \vee y \le z$ then $(x \vee y) \to z = 1$ and also we have $x \le z$ and $y \le z$. Then $(x \to z) \wedge (y \to z) = 1 \wedge 1 = 1$.

Case 2. If $x \vee y > z$ then $(x \vee y) \to z = z$ and $x \ge z$ or $,y \ge z$ so we have $(x \to z) = z$ or $(y \to z) = z$ and then $(x \to z) \wedge (y \to z) = z$.

(ii) Case 1. If $x \wedge y \le z$ then $(x \wedge y) \to z = 1$. Also, we have either $x \le z$ or $y \le z$ in which case either $x \to z = 1$ or $y \to z = 1$ and so $(x \to z) \vee (y \to z) = 1$.

Case 2. If $x \wedge y > z$, then $(x \wedge y) \to z = z$. Also, if $(x \wedge y) > z$ then both $x > z$ and $y > z$ and $x \to z = z$ and $y \to z = z$.

(iii) Case 1. If $x \le (y \vee z)$ then $x \to (y \vee z) = 1$. Also we have either $x \le y$ or $x \le z$ and so $x \to y = 1$ or $x \to z = 1$.

Case 2. Otherwise $x \to (y \vee z) = y \vee z$ and we have $x > y$ and $x > z$ and $x \to y = y$ and $x \to z = z$. Then $(x \to y) \vee (x \to z) = y \vee z$.

(iv) The proof is left as an exercise to the reader.

(v) Case 1. If $x \le y$ then $x \wedge (x \to y) = x \wedge 1 = x \le y$.
Case 2. Otherwise $x \wedge (x \to y) = x \wedge y = y$.

(vi) Case 1. If $x \le y$ then $x \to (x \wedge y) = x \to x = 1 \ge y$.
Case 2. Otherwise when $x > y$ then we have $x \to (x \wedge y) = x \to y = y$.

(vii) The proof is left as an exercise to the reader.    ∎

**Definition 3.19.** *The **min→** composition can be defined as:*

$$R \triangleleft S(x, z) = \bigwedge_{y \in Y} R(x, y) \to S(y, z).$$

Often in the literature (see e.g. De Baets [42]) it is called subcomposition and a dual operation is considered as

$$R \triangleright S(x, z) = \bigwedge_{y \in Y} S(y, z) \to R(x, y)$$

called the supercomposition. The relation between the two is given by the next proposition.

**Proposition 3.20.** *For $R_1 \in \mathcal{F}(X \times Y)$ and $R_2 \in \mathcal{F}(Y \times Z)$ we have*
(i) $R_1 \lhd R_2 = (R_2^{-1} \rhd R_1^{-1})^{-1}$;
(ii) $R_1 \rhd R_2 = (R_2^{-1} \lhd R_1^{-1})^{-1}$.

**Proof.** The proof is left to the reader as an exercise. ∎

In the present work we will mainly use the subcomposition and we call it simply min→ composition.

**Proposition 3.21.** *If $R, S \in \mathcal{F}(X \times Y)$ and $Q \in \mathcal{F}(Y \times Z)$ are such that $R \le S$ then $R \lhd Q \ge S \lhd Q$.*

**Proof.** Since → is decreasing in the first argument we have

$$R \lhd Q(x, z) = \bigwedge_{y \in Y} R(x, y) \to Q(y, z)$$

$$\ge \bigwedge_{y \in Y} S(x, y) \to Q(y, z) = R \lhd S(x, z).$$

∎

**Proposition 3.22.** *For any $R, S \in \mathcal{F}(X \times Y)$ and $Q \in \mathcal{F}(Y \times Z)$ we have*
(i) $(R \vee S) \lhd Q = (R \lhd Q) \wedge (S \lhd Q)$.
(ii) $(R \wedge S) \lhd Q \ge (R \lhd Q) \vee (S \lhd Q)$.

**Proof.** (i) From the previous proposition we have $(R \vee S) \lhd Q \le R \lhd Q$ and $(R \vee S) \lhd Q \le S \lhd Q$ and then

$$(R \vee S) \lhd Q \le (R \lhd Q) \wedge (S \lhd Q).$$

Also, from Proposition 3.18, (i)

$$(R \vee S) \lhd Q(x, z) = \bigwedge_{y \in Y} (R(x, y) \vee S(x, y)) \to Q(y, z)$$

$$= \bigwedge_{y \in Y} (R(x, y) \to Q(y, z)) \wedge (S(x, y) \to Q(y, z))$$

$$\ge \bigwedge_{y \in Y} (R(x, y) \to Q(y, z)) \wedge \bigwedge_{y \in Y} (S(x, y) \to Q(y, z))$$

$$= (R \lhd Q) \wedge (S \lhd Q)(x, z).$$

(ii) Similar to (i) we have $(R \wedge S) \lhd Q \ge R \lhd Q$ and $(R \wedge S) \lhd Q \ge S \lhd Q$ and then

$$(R \wedge S) \lhd Q \ge (R \lhd Q) \vee (S \lhd Q).$$

∎

## 3.5    Fuzzy Relational Equations with Max-Min and Min $\to$ Compositions

We consider the following two fuzzy relational equations

$$R \circ P = Q$$

and

$$R \triangleleft P = Q$$

with $R \in \mathcal{F}(X \times Y)$, $P \in \mathcal{F}(Y \times Z)$ and $Q \in \mathcal{F}(X \times Z)$.

**Theorem 3.23.** *The following inequalities hold true:*
*(i)* $P \leq R^{-1} \triangleleft (R \circ P)$;
*(ii)* $R \circ (R^{-1} \triangleleft Q) \leq Q$;
*(iii)* $R \leq (P \triangleleft (R \circ P)^{-1})^{-1}$;
*(iv)* $(P \triangleleft Q^{-1})^{-1} \circ P \leq Q$.

**Proof.** (i) Using Proposition 3.18, (vi) we have for every $y \in Y$ and $z \in Z$,

$$R^{-1} \triangleleft (R \circ P)(y, z) = \bigwedge_{x \in X} R^{-1}(y, x) \to (R \circ P)(x, z)$$

$$= \bigwedge_{x \in X} R^{-1}(y, x) \to \bigvee_{t \in Y} R(x, t) \wedge P(t, z)$$

$$\geq \bigwedge_{x \in X} R(x, y) \to R(x, y) \wedge P(y, z) \geq \bigwedge_{x \in X} P(y, z) = P(y, z).$$

(ii) From Proposition 3.18, (v) we obtain for every $x \in X$ and $z \in Z$

$$R \circ (R^{-1} \triangleleft Q)(x, z) = \bigvee_{y \in Y} R(x, y) \wedge (R^{-1} \triangleleft Q)(y, z)$$

$$= \bigvee_{y \in Y} R(x, y) \wedge (\bigwedge_{t \in X} R^{-1}(y, t) \to Q(t, z))$$

$$\leq \bigvee_{y \in Y} R(x, y) \wedge (R(x, y) \to Q(x, z)) \leq \bigvee_{y \in Y} Q(x, z) = Q(x, z).$$

(iii) We have for every $x \in X$ and $y \in Y$

$$(P \triangleleft (R \circ P)^{-1})^{-1}(x, y) = P \triangleleft (R \circ P)^{-1}(y, x)$$

$$= \bigwedge_{z \in Z} P(y, z) \to (R \circ P)^{-1}(z, x)$$

$$= \bigwedge_{z \in Z} P(y, z) \to \bigvee_{t \in Y} R(x, t) \wedge P(t, z)$$

$$\geq \bigwedge_{z \in Z} P(y, z) \to R(x, y) \wedge P(y, z) \geq \bigwedge_{z \in Z} R(x, y) = R(x, y).$$

(iv) The proof of (iv) is similar to that of (iii), and it uses Proposition 3.18. ∎

**Theorem 3.24.** *(Sanchez [129]) (i) Consider the equation $R \circ P = Q$ with unknown $P$. The equation has solutions if and only if $R^{-1} \triangleleft Q$ is a solution and in this case it is the greatest solution of this equation.*

*(ii) Consider the equation $R \circ P = Q$ with unknown $R$. The equation has solutions if and only if $(P \triangleleft Q^{-1})^{-1}$ is a solution and in this case it is the greatest solution of this equation.*

**Proof.** (i) "⟸" If $P = R^{-1} \triangleleft Q$ is a solution of the equation $R \circ P = Q$ then the equation possesses solutions.

"⟹" Let us suppose now that the equation $R \circ P = Q$ has a solution denoted by $P$. Then by Theorem 3.23, (i) we have

$$P \leq R^{-1} \triangleleft (R \circ P) = R^{-1} \triangleleft Q,$$

i.e., $R^{-1} \triangleleft Q$ is greater than or equal to any solution $P$ of the equation $R \circ P = Q$.

To show that $R^{-1} \triangleleft Q$ is a solution we use double inclusion. By Theorem 3.23, (ii) we know

$$R \circ (R^{-1} \triangleleft Q) \leq Q.$$

On the other hand since $P \leq R^{-1} \triangleleft Q$ and since $\circ$ is increasing in its second argument, we have

$$R \circ (R^{-1} \triangleleft Q) \geq R \circ P = Q.$$

Combining the two inequalities we get $R \circ (R^{-1} \triangleleft Q) = Q$, i.e., $R^{-1} \triangleleft Q$ is a solution and moreover, it is the greatest solution.

(ii) The proof of (ii) is similar. ∎

**Example 3.25.** *Let us consider the fuzzy relational equation*

$$\begin{pmatrix} 0.3 & 0.2 & 0.4 \\ 0.1 & 0.3 & 0.5 \\ 0.5 & 0.4 & 0.6 \end{pmatrix} \circ P = \begin{pmatrix} 0.3 & 0.4 & 0.4 \\ 0.3 & 0.5 & 0.5 \\ 0.3 & 0.5 & 0.6 \end{pmatrix}.$$

*Then*

$$R^{-1} \triangleleft Q = \begin{pmatrix} 0.3 & 1 & 1 \\ 0.3 & 1 & 1 \\ 0.3 & 0.5 & 1 \end{pmatrix},$$

*and since*

$$
\begin{pmatrix} 0.3 & 0.2 & 0.4 \\ 0.1 & 0.3 & 0.5 \\ 0.5 & 0.4 & 0.6 \end{pmatrix} \circ \begin{pmatrix} 0.3 & 1 & 1 \\ 0.3 & 1 & 1 \\ 0.3 & 0.5 & 1 \end{pmatrix} = \begin{pmatrix} 0.3 & 0.4 & 0.4 \\ 0.3 & 0.5 & 0.5 \\ 0.3 & 0.5 & 0.6 \end{pmatrix},
$$

then $R^{-1} \triangleleft Q$ is a solution of the equation. From the previous theorem it follows that it is the greatest solution of the given equation.

**Theorem 3.26.** *The following inequalities hold true:*
  *(i)* $(Q \triangleleft P^{-1}) \triangleleft P \geq Q;$
  *(ii)* $(R \triangleleft P) \triangleleft P^{-1} \geq R;$
  *(iii)* $R^{-1} \circ (R \triangleleft P) \leq P;$
  *(iv)* $R \triangleleft (R^{-1} \circ Q) \geq Q.$

**Proof.** (i) Using Proposition 3.18, (vii), and taking into account that $\rightarrow$ is decreasing in its first argument we have

$$
(Q \triangleleft P^{-1}) \triangleleft P(x, z) = \bigwedge_{y \in Y} (Q \triangleleft P^{-1})(x, y) \rightarrow P(y, z)
$$

$$
= \bigwedge_{y \in Y} (\bigwedge_{t \in Z} Q(x, t) \rightarrow P(y, t)) \rightarrow P(y, z)
$$

$$
\geq \bigwedge_{y \in Y} (Q(x, z) \rightarrow P(y, z)) \rightarrow P(y, z) \geq \bigwedge_{y \in Y} Q(x, z) = Q(x, z).
$$

  (ii), (iii), (iv) The proofs of these properties are similar and they are using the results in Proposition 3.18                                              ∎

**Theorem 3.27.** *(Miyakoshi-Shimbo [112]) (i) Consider the equation $R \triangleleft P = Q$ with unknown R. The equation has solutions if and only if $Q \triangleleft P^{-1}$ is a solution and in this case it is the greatest solution of this equation.*
  *(ii) Consider the equation $R \triangleleft P = Q$ with unknown P. The equation has solutions if and only if $R^{-1} \circ Q$ is a solution and in this case it is the least solution of this equation.*

**Proof.** (i) "$\Leftarrow$" Obvious.
  "$\Rightarrow$" Let us suppose now that the equation $R \triangleleft P = Q$ has solution R. Then by Theorem 3.26, (ii) we have

$$
R \leq (R \triangleleft P) \triangleleft P^{-1} = Q \triangleleft P^{-1}.
$$

i.e., $Q \triangleleft P^{-1}$ is greater then any solution R.
  By Theorem 3.26, (i) we know that

$$
Q \leq (Q \triangleleft P^{-1}) \triangleleft P.
$$

On the other hand since $R \leq Q \vartriangleleft P^{-1}$ and since $\vartriangleleft$ is decreasing in its first argument, we have

$$Q = R \vartriangleleft P \geq (Q \vartriangleleft P^{-1}) \vartriangleleft P.$$

Combining the two inequalities we get $Q = (Q \vartriangleleft P^{-1}) \vartriangleleft P$.

(ii) "$\Leftarrow$" Obvious.

"$\Rightarrow$" Let $P$ be a solution of $R \vartriangleleft P = Q$. Then by Theorem 3.26, (iii),

$$P \geq R^{-1} \circ (R \vartriangleleft P) = R^{-1} \circ Q,$$

i.e., $R^{-1} \circ Q$ is less then any solution $P$.

Also, by Theorem 3.26, (iv)

$$Q \leq R \vartriangleleft (R^{-1} \circ Q)$$

since $P \geq R^{-1} \vartriangleleft Q$ and taking into account that $\vartriangleleft$ is increasing in its second argument we have

$$Q = R \vartriangleleft P \geq R \vartriangleleft (R^{-1} \circ Q)$$

and we get $Q = R \vartriangleleft (R^{-1} \circ Q)$. ∎

**Example 3.28.** *Let us consider the fuzzy relational equation*

$$\begin{pmatrix} 0.3 & 0.2 & 0.4 \\ 0.1 & 0.3 & 0.5 \\ 0.5 & 0.4 & 0.6 \end{pmatrix} \vartriangleleft P = \begin{pmatrix} 0.3 & 0.4 & 0.4 \\ 1 & 0.6 & 0.5 \\ 0.5 & 0.5 & 0.6 \end{pmatrix}.$$

*Then*

$$R^{-1} \circ Q = \begin{pmatrix} 0.5 & 0.5 & 0.5 \\ 0.4 & 0.4 & 0.4 \\ 0.5 & 0.5 & 0.6 \end{pmatrix}.$$

*Since*

$$\begin{pmatrix} 0.3 & 0.2 & 0.4 \\ 0.1 & 0.3 & 0.5 \\ 0.5 & 0.4 & 0.6 \end{pmatrix} \vartriangleleft \begin{pmatrix} 0.5 & 0.5 & 0.5 \\ 0.4 & 0.4 & 0.4 \\ 0.5 & 0.5 & 0.6 \end{pmatrix} = \begin{pmatrix} 1 & 1 & 1 \\ 1 & 1 & 1 \\ 0.5 & 0.5 & 1 \end{pmatrix}.$$

*is not a solution of the equation, by the preceding theorem it follows that the equation has no solutions.*

## 3.6 Max-t-Norm Composition

**Definition 3.29.** *The max-min composition can be naturally generalized to max-t-norm compositions*

$$R \circ_T P(x, z) = \bigvee_{y \in Y} R(x, y) \, T \, P(y, z),$$

*where $T$ is an arbitrary t-norm.*

**Proposition 3.30.** *(i) The max-t-norm composition is associative, i.e.,*

$$(R \circ_T S) \circ_T Q = R \circ_T (S \circ_T Q),$$

*for any* $R \in \mathcal{F}(X \times Y)$, $S \in \mathcal{F}(Y \times Z)$ *and* $Q \in \mathcal{F}(Z \times U)$.
 *(ii) If* $R_1 \leq R_2$ *then*

$$R_1 \circ_T Q \leq R_2 \circ_T Q$$

*for any* $R_1, R_2 \in \mathcal{F}(X \times Y)$ *and* $Q \in \mathcal{F}(Y \times Z)$.

**Proof.** Since the t-norm $T$ is increasing we have

$$(R \circ_T S) \circ_T Q(x, u) = \bigvee_{z \in Z} \left( \bigvee_{y \in Y} R(x, y) \; T \; S(y, z) \right) T \; Q(z, u)$$

$$= \bigvee_{z \in Z} \bigvee_{y \in Y} R(x, y) \; T \; S(y, z) \; T \; Q(z, u) = R \circ_T (S \circ_T Q)(x, u).$$

(ii) The proof of (ii) is based on the fact that $T$ is increasing in both of its variables.  ∎

**Proposition 3.31.** *For any* $R, S \in \mathcal{F}(X \times Y)$, $Q \in \mathcal{F}(Y \times Z)$ *and any t-norm* $T$ *we have*
 *(i)* $(R \vee S) \circ_T Q = (R \circ_T Q) \vee (S \circ_T Q)$
 *(ii)* $(R \wedge S) \circ_T Q \leq (R \circ_T Q) \wedge (S \circ_T Q)$.

**Proof.** (i) Since $T$ is distributive with respect to $\vee$ we have

$$(R \vee S) \circ_T Q(x, z) = \bigvee_{y \in Y} (R \vee S)(x, y) \; T \; Q(y, z)$$

$$\leq \bigvee_{y \in Y} (R(x, y) \; T \; Q(y, z)) \vee \bigvee_{y \in Y} (S(x, y) \; T \; Q(y, z)),$$

Also, $R \circ_T Q \leq (R \vee S) \circ_T Q$ and $S \circ_T Q \leq (R \vee S) \circ_T Q$, and then

$$(R \circ_T Q) \vee (S \circ_T Q) = (R \vee S) \circ_T Q.$$

(ii) Is left to the reader as an exercise.  ∎

## 3.7   Min $\to_T$ Composition

Let $T$ be an arbitrary continuous t-norm and $\to_T$ be the R-implication defined as

$$x \to_T y = \sup\{z | x \; T \; z \leq y\}.$$

**Proposition 3.32.** *For any $x, y, z \in [0, 1]$ and for any t-norm $T$ the residual implication $\to_T$ has the following properties:*

*(i) $xT(x \to_T y) \leq y$.*
*(ii) $x \to_T (xTy) \geq y$.*
*(iii) $(x \to_T y) \to_T y \geq x$.*

**Proof.** (i) We have

$$xT(x \to_T y) = xT \sup\{z|xTz \leq y\}.$$

Let now $\varepsilon > 0$ arbitrary. Then there exist a $z_0 \in \{z|xTz \leq y\}$ such that $x \to_T y < z_0 + \varepsilon$. Then

$$xT(x \to_T y) < xT(z_0 + \varepsilon).$$

Since $T$ is continuous this implies

$$xT(x \to_T y) \leq xTz_0.$$

Now, the condition $z_0 \in \{z|xTz \leq y\}$ implies $xTz_0 \leq y$, and then we get $xT(x \to_T y) \leq y$.

(ii) It is easy to observe that $y \in \{z|xTz \leq xTy\}$ and so,

$$x \to_T (xTy) = \sup\{z|xTz \leq xTy\} \geq y.$$

(iii) To prove the last inequality we observe that

$$(x \to_T y) \to_T y = \sup\{z|(x \to_T y)Tz \leq y\}.$$

Using now (i) we have $(x \to_T y)Tx \leq y$ and then $x \in \{z|(x \to_T y)Tz \leq y\}$. Then $(x \to_T y) \to_T y \geq x$. ∎

**Definition 3.33.** *The **min$\to_T$ composition** can be defined in a similar way as the min $\to$ composition:*

$$R \triangleleft_T S(x, z) = \bigwedge_{y \in Y} R(x, y) \to_T S(y, z)$$

**Proposition 3.34.** *If $R, S \in \mathcal{F}(X \times Y)$ are such that $R \leq S$ and if $Q \in \mathcal{F}(Y \times Z)$ then $R \triangleleft_T Q \geq S \triangleleft Q$.*

**Proof.** The proof is left to the reader. ∎

# 3.8 Fuzzy Relational Equations with Max-t-Norm and Min $\to_T$ Compositions

Fuzzy relational equations with max-t-norm or min $\to_T$ compositions can be studied in a similar way as the equations above.

We consider the following two fuzzy relational equations with max-t-norm and min $\to_T$ compositions

$$R \circ_T P = Q$$

and

$$R \triangleleft_T P = Q$$

with $R \in \mathcal{F}(X \times Y)$, $P \in \mathcal{F}(Y \times Z)$ and $Q \in \mathcal{F}(X \times Z)$.

Theorem 3.23 can be generalized as follows.

**Theorem 3.35.** *The following inequalities hold true:*
*(i) $P \leq R^{-1} \triangleleft_T (R \circ_T P)$;*
*(ii) $R \circ_T (R^{-1} \triangleleft_T Q) \leq Q$;*
*(iii) $R \leq (P \triangleleft_T (R \circ_T P)^{-1})^{-1}$;*
*(iv) $(P \triangleleft_T Q^{-1})^{-1} \circ_T P \leq Q$.*

**Proof.** First let us observe that since $\to_T$ is a fuzzy implication, it is decreasing in the first argument and increasing in the second argument.

(i) Using Proposition 3.32, (ii) we have for $y \in Y$ and $z \in Z$,

$$R^{-1} \triangleleft_T (R \circ_T P)(y,z) = \bigwedge_{x \in X} R^{-1}(y,x) \to_T \bigvee_{t \in Y} R(x,t)TP(t,z) \geq$$

$$\geq \bigwedge_{x \in X} R(x,y) \to_T (R(x,y)TP(y,z)) \geq \bigwedge_{x \in X} P(y,z) = P(y,z).$$

(ii) From Proposition 3.18, (i) we obtain

$$R \circ_T (R^{-1} \triangleleft_T Q)(x,z) = \bigvee_{y \in Y} R(x,y)T(\bigwedge_{t \in X} R^{-1}(y,t) \to_T Q(t,z)) \leq$$

$$\leq \bigvee_{y \in Y} R(x,y)T(R(x,y) \to_T Q(x,z)) \leq \bigvee_{y \in Y} Q(x,z) = Q(x,z).$$

(iii), (iv) The proofs of (iii) and (iv) are similar to those of (i), (ii).    ∎

**Theorem 3.36.** *(Sanchez [129], Miyakoshi-Shimbo [112]) (i) Consider the equation $R \circ_T P = Q$ with unknown $P$. The equation has solutions if and only if $R^{-1} \triangleleft_T Q$ is a solution and in this case it is the greatest solution of this equation.*

*(ii) Consider the equation $R \circ_T P = Q$ with unknown $R$. The equation has solutions if and only if $(P \triangleleft Q^{-1})^{-1}$ is a solution and in this case it is the greatest solution of this equation.*

**Proof.** The proof of this theorem is similar to that of Theorem 3.24.    ∎

**Theorem 3.37.** *The following inequalities hold true:*
(i) $(Q \triangleleft_T P^{-1}) \triangleleft_T P \geq Q;$
(ii) $(R \triangleleft_T P) \triangleleft_T P^{-1} \geq R;$
(iii) $R^{-1} \circ_T (R \triangleleft_T P) \leq P;$
(iv) $R \triangleleft_T (R^{-1} \circ_T Q) \geq Q.$

**Proof.** (i) Using Proposition 3.32, (iii), and taking into account that $\to_T$ is decreasing in its first argument we have

$$(Q \triangleleft_T P^{-1}) \triangleleft_T P(x,z) = \bigwedge_{y \in Y} \left( \bigwedge_{t \in Z} Q(x,t) \to_T P(y,t) \right) \to_T P(y,z) \geq$$

$$\geq \bigwedge_{y \in Y} (Q(x,z) \to_T P(y,z)) \to_T P(y,z) \geq \bigwedge_{y \in Y} Q(x,z) = Q(x,z).$$

(ii), (iii), (iv) The proofs of these properties are similar to the previous reasoning. ∎

**Theorem 3.38.** *(Miyakoshi-Shimbo [112]) (i) Consider the equation $R \triangleleft_T P = Q$ with unknown R. The equation has solutions if and only if $Q \triangleleft_T P^{-1}$ is a solution and in this case it is the greatest solution of this equation.*
*(ii) Consider the equation $R \triangleleft_T P = Q$ with unknown P. The equation has solutions if and only if $R^{-1} \circ_T Q$ is a solution and in this case it is the least solution of this equation.*

**Proof.** The proof follows the ideas in the proof of Miyakoshi-Shimbo Theorem 3.27. ∎

## 3.9 Problems

1. Consider the fuzzy relations

$$R = \begin{pmatrix} 0.1 & 0.4 & 0.2 & 0.5 \\ 0 & 0.2 & 0.3 & 0.4 \\ 0.4 & 0.5 & 0.2 & 0.1 \\ 1 & 0.3 & 0.3 & 0 \end{pmatrix}$$

and

$$Q = \begin{pmatrix} 0.3 & 0.3 & 0.2 & 0.1 \\ 0.6 & 0.5 & 0 & 0.2 \\ 0.2 & 0.1 & 0.4 & 0.1 \\ 0.4 & 0.1 & 0.2 & 0.5 \end{pmatrix}.$$

Calculate $R \circ Q$, $R \bullet Q$, $R \triangleleft Q$, $R \triangleright Q$.

2. Given

$$R = \begin{pmatrix} 0.1 & 0.4 & 0.2 & 0.4 \\ 0.4 & 0.5 & 0.2 & 0.1 \\ 0.8 & 0.3 & 0.6 & 0.2 \end{pmatrix}$$

and

$$Q = \begin{pmatrix} 0.3 & 0.1 & 0.2 \\ 0.5 & 0.1 & 0.5 \\ 0.3 & 0.2 & 0.6 \end{pmatrix}$$

decide whether the equation $R \circ P = Q$ does or does not have a solution. If the answer is yes, find the greatest solution.

3. Given

$$R = \begin{pmatrix} 0.1 & 0.4 & 0.2 & 0.4 \\ 0.4 & 0.5 & 0.2 & 0.1 \\ 0.8 & 0.3 & 0.6 & 0.2 \end{pmatrix}$$

and

$$Q = \begin{pmatrix} 0.3 & 0.1 & 0.2 \\ 0.5 & 0.1 & 0.5 \\ 0.3 & 0.2 & 0.6 \end{pmatrix}$$

decide whether the equation $R \triangleleft P = Q$ does or does not have a solution. If the answer is yes, find the least solution.

4. Prove Proposition 3.18, (iv) and (vii)

5. Prove Proposition 3.20.

6. Prove Theorem 3.23, (iv)

7. Prove Theorem 3.24, (ii)

8. Prove Theorem 3.26, (ii), (iii), (iv).

9. Prove Proposition 3.31.

10. Prove Proposition 3.34.

11. Consider the fuzzy relational equation $R \bullet P = Q$. Find a solvability condition and find a solution of this equation, supposing that it is solvable.

12. Let $R \in \mathcal{F}(X \times X)$ be a fuzzy relation. Let us consider the eigen fuzzy sets equation $R \circ A = A$, with $A \in \mathcal{F}(X)$. Prove that the equation has a least and a greatest solution and these can be calculated as

$$A_l = \lim_{n \to \infty} R^n \circ \emptyset.$$

and

$$A_g = \lim_{n \to \infty} R^n \circ X.$$

Hint: Use Tarski-Knaster and Kleene's fixed point theorems (see appendix).

13. Let $R \in \mathcal{F}(X \times X)$ be a fuzzy relation. Let us consider the eigen fuzzy sets equation $R \bullet A = A$, with $A \in \mathcal{F}(X)$. Prove that the equation has a least and a greatest solution.

14. Given $R \in \mathcal{F}(X \times X)$ we consider the convex combined composition defined as
$$R * A = \lambda(R \circ A) + (1 - \lambda)(R \bullet A),$$

    with $A \in \mathcal{F}(X)$. Prove that the equation has a least and a greatest solution.

15. Generalize problems 12, 13, 14 to the case of max-t-norm composition $R \circ_T A$, min-t-conorm $R \bullet_S A$ and convex combined composition
$$R *_T A = \lambda(R \circ_T A) + (1 - \lambda)(R \bullet_T A).$$

16. Let $R \in \mathcal{F}(X \times X)$ be a fuzzy relation. Consider the eigen fuzzy sets equation with min $\to$ composition $R \triangleleft A = A$, $A \in \mathcal{F}(X)$. Investigate existence and the structure of solutions of this equation.

# 4
# Fuzzy Numbers

## 4.1 Definition of Fuzzy Numbers

Fuzzy numbers generalize classical real numbers and roughly speaking a fuzzy number is a fuzzy subset of the real line that has some additional properties. Fuzzy numbers are capable of modeling epistemic uncertainty and its propagation through calculations. The fuzzy number concept is basic for fuzzy analysis and fuzzy differential equations, and a very useful tool in several applications of fuzzy sets and fuzzy logic.

**Definition 4.1.** *(see e.g., Diamond-Kloeden [44]) Consider a fuzzy subset of the real line $u : \mathbb{R} \to [0,1]$. Then $u$ is a **fuzzy number** if it satisfies the following properties:*

*(i) $u$ is normal, i.e. $\exists x_0 \in \mathbb{R}$ with $u(x_0) = 1$;*

*(ii) $u$ is fuzzy convex (i.e. $u(tx + (1-t)y) \geq \min\{u(x), u(y)\}$, $\forall t \in [0,1]$, $x, y \in \mathbb{R}$);*

*(iii) $u$ is upper semicontinuous on $\mathbb{R}$ (i.e. $\forall \varepsilon > 0 \; \exists \delta > 0$ such that $u(x) - u(x_0) < \varepsilon$, $|x - x_0| < \delta$).*

*(iv) $u$ is compactly supported i.e., $cl\{x \in \mathbb{R}; \; u(x) > 0\}$ is compact, where $cl(A)$ denotes the closure of the set $A$.*

*Let us denote by $\mathbb{R}_{\mathcal{F}}$ the **space of fuzzy numbers**.*

**Example 4.2.** *The fuzzy set $u : \mathbb{R} \to [0,1]$,*

$$u(x) = \begin{cases} 0 & \text{if} & x < 0 \\ x^3 & \text{if} & 0 \leq x < 1 \\ (2-x)^3 & \text{if} & 1 \leq x \leq 2 \\ 0 & \text{if} & x \geq 2 \end{cases}$$

*is a fuzzy number. See figure 4.1.*

B. Bede: *Mathematics of Fuzzy Sets and Fuzzy Logic*, STUDFUZZ 295, pp. 51–64.
DOI: 10.1007/978-3-642-35221-8_4    © Springer-Verlag Berlin Heidelberg 2013

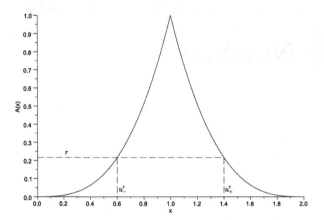

**Fig. 4.1** Example of a fuzzy number and its level sets

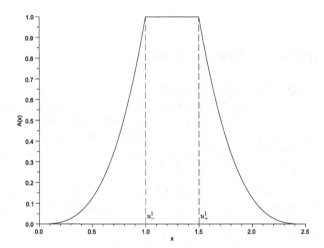

**Fig. 4.2** Example of a fuzzy number and its core

**Example 4.3.** *The fuzzy set*

$$
u(x) = \begin{cases}
0 & \text{if} & x < 0 \\
x^3 & \text{if} & 0 \le x \le 1 \\
1 & \text{if} & 1 \le x \le 1.5 \\
(2.5 - x)^3 & \text{if} & 1.5 < x \le 2.5 \\
0 & \text{if} & x > 2.5
\end{cases}
$$

*in Figure 4.2 is a fuzzy number as well.*

**Example 4.4.** *The fuzzy set represented in Figure 4.3 is not a fuzzy number since it is not fuzzy convex.*

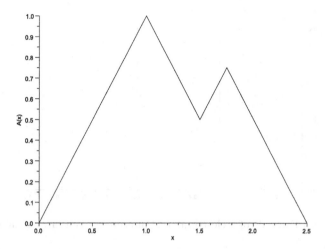

**Fig. 4.3** Example of a fuzzy set that is not a fuzzy number

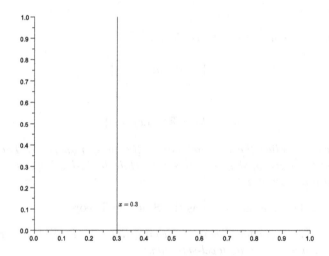

**Fig. 4.4** A singleton fuzzy number

**Remark 4.5.** *Obviously any real number is also a fuzzy number, i.e.,* $\mathbb{R} \subset \mathbb{R}_{\mathcal{F}}$. *Here* $\mathbb{R} \subset \mathbb{R}_{\mathcal{F}}$ *is understood as*

$$\mathbb{R} = \left\{ \chi_{\{x\}}; x \text{ is usual real number} \right\}.$$

$\chi_{\{x\}}$ *is a singleton fuzzy number for any given real number* $x \in \mathbb{R}$ *and it can be identified with* $x \in \mathbb{R}$ *(see Figure 4.4)*

Also, fuzzy numbers generalize closed intervals. Indeed, if $\mathbb{I}$ denotes the set of all real intervals, then $\mathbb{I} \subset \mathbb{R}_{\mathcal{F}}$, where

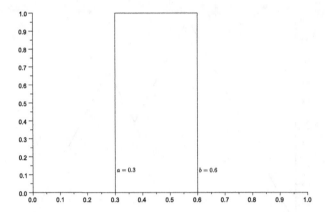

**Fig. 4.5** Example of a closed interval interpreted as a fuzzy number

$$\mathbb{I} = \left\{\chi_{[a,b]}; [a,b] \text{ is a usual real interval}\right\}$$

(see Figure 4.5).

**Definition 4.6.** *For $0 < r \leq 1$, we denote*

$$u_r = \{x \in \mathbb{R}; u(x) \geq r\}$$

*and*

$$u_0 = cl\{x \in \mathbb{R};\ u(x) > 0\}.$$

*Then $u_r$ will be called the $r-$**level set** of the fuzzy number $u$. The 1-level set is called the **core** of the fuzzy number, while the 0-level set is called the **support** of the fuzzy number.*

The following Theorem is known as the Stacking Theorem.

**Theorem 4.7.** *(Stacking Theorem, Negoita-Ralescu [115]) If $u \in \mathbb{R}_{\mathcal{F}}$ is a fuzzy number and $u_r$ are its level-sets then:*
*(i) $u_r$ is a closed interval $u_r = [u_r^-, u_r^+]$, for any $r \in [0,1]$;*
*(ii) If $0 \leq r_1 \leq r_2 \leq 1$, then $u_{r_2} \subseteq u_{r_1}$.*
*(iii) For any sequence $r_n$ which converges from below to $r \in (0,1]$ we have*

$$\bigcap_{n=1}^{\infty} u_{r_n} = u_r.$$

*(iv) For any sequence $r_n$ which converges from above to $0$ we have*

$$cl\left(\bigcup_{n=1}^{\infty} u_{r_n}\right) = u_0.$$

**Proof.** (i) Let us first remark that all the sets $u_r$ are nonempty and bounded since $u_1 \neq \emptyset$ and because $u_0$ is bounded (a compact set in $\mathbb{R}$ is closed and bounded). Let $u$ be a fuzzy number and $r \in (0, 1]$. If $a, b \in u_r$, then $u(a) \geq r$ and $u(b) \geq r$. Then from the fuzzy convexity, if $x \in [a, b]$ we get

$$u(x) \geq \min\{u(a), u(b)\} = r,$$

i.e., $x \in [a, b]$. As a conclusion $u$ contains any closed interval $[a, b]$ together with the points $[a, b]$ this means that $u_r$ is convex. All is left to be proven is that $u_r$ is closed. Upper semicontinuity implies that if $u(x_0) < r$ then there is an open interval $W$ with $x_0 \in W$ such that $u(x) < r, \forall x \in W$. It immediately follows that the set $\{x | u(x) < r\}$ is open and then it has a closed complement, i.e., $u_r$ is closed. On the real line, closed convex sets are closed intervals so $u_r$ is a closed interval for any $r \in [0, 1]$.

(ii) It is easy to check that (ii) holds. Indeed, if $0 < r_1 \leq r_2 \leq 1$ then if $x \in u_{r_2}$ then $u(x) \geq r_2 \geq r_1$ and so, $x \in u_{r_1}$. If $r_1 = 0$ or $r_2 = 0$ then the proof of (ii) is immediate.

(iii) Let $r_n \nearrow r$ be non-decreasing. Then $u_{r_n} \subseteq u_{r_{n-1}}$, is a descending sequence of closed intervals $u_{r_n} = [u_{r_n}^-, u_{r_n}^+]$. Then it is easy to see that $u_{r_n}^-, u_{r_n}^+$ converge, in which case let $u_{r_n}^- \to a$, $u_{r_n}^+ \to b$ and

$$[a, b] = \bigcap_{n=1}^{\infty} u_{r_n}.$$

We have to prove that $u_r = [a, b]$. For a given $x \in [a, b]$ we have

$$u(x) \geq \min\{u(a), u(b)\}.$$

So, it is enough to show that $u(a) \geq r$ and $u(b) \geq r$. Suppose that $u(a) < r$, then since $u$ is upper semicontinuous, there is a neighborhood $W$ of $a$, such that $u(x) < r$. This implies the existence of a rank $N \in \mathbb{N}$ with $u(u_{r_n}^-) < r$ for any $n \geq N$. Then since $r_n \to r$ we obtain that there exists $n \in \mathbb{N}$ such that $u(u_{r_n}^-) < r_n$ which is a contradiction. Then it follows that $u(a) \geq r$. Similarly we can show that $u(b) \geq r$ so, $u(x) \geq r$ and then $[a, b] \subseteq u_r$. Form (ii), we have $u_r \subseteq u_{r_n}$ and it implies $u_r \subseteq [a, b]$. Then finally we get $u_r = [a, b]$, i.e.,

$$\bigcap_{n=1}^{\infty} u_{r_n} = u_r.$$

(iv) For (iv) we observe that the inclusion

$$\bigcup_{n=1}^{\infty} u_{r_n} \subseteq u_0$$

is straightforward. Since $u_0$ is closed we get

$$cl\left(\bigcup_{n=1}^{\infty} u_{r_n}\right) \subseteq u_0$$

Reciprocally, $x \in u_0$ implies that there is a sequence

$$x_n \in \{x \in \mathbb{R};\ u(x) > 0\}$$

that converges to $x$. Without loss of generality we may assume that

$$x_n \in u_{r_n} \subseteq \bigcup_{n=1}^{\infty} u_{r_n}.$$

Then we obtain

$$x \in cl\left(\bigcup_{n=1}^{\infty} u_{r_n}\right).$$

The two inclusions lead to the required conclusion.    ∎

The endpoints of the $r$-level set $u_r$ are given by $u_r^- = \inf u_r$ and $u_r^+ = \sup u_r$. Then also $u_r = [u_r^-, u_r^+]$. (See Figure 4.1).

## 4.2   Characterization Theorems for Fuzzy Numbers

The following theorem is Negoita-Ralescu characterization theorem, and it is the reciprocal of the stacking theorem.

**Theorem   4.8.** *(Negoita-Ralescu   characterization   theorem,   Negoita-Ralescu [115]) Given a family of subsets $\{M_r : r \in [0,1]\}$ that satisfies conditions (i)-(iv)*
*(i) $M_r$ is a non-empty closed interval for any $r \in [0,1]$;*
*(ii) If $0 \leq r_1 \leq r_2 \leq 1$, we have $M_{r_2} \subseteq M_{r_1}$;*
*(iii) For any sequence $r_n$ which converges from below to $r \in (0,1]$ we have*

$$\bigcap_{n=1}^{\infty} M_{r_n} = M_r;$$

*(iv) For any sequence $r_n$ which converges from above to $0$ we have*

$$cl\left(\bigcup_{n=1}^{\infty} M_{r_n}\right) = M_0.$$

*Then there exists a unique $u \in \mathbb{R}_{\mathcal{F}}$, such that $u_r = M_r$, for any $r \in [0,1]$.*

**Proof.** Let $M_r$ fulfill the properties (i)-(iv). Then by defining

$$u(x) = \begin{cases} 0 & \text{if } x \notin M_0 \\ \sup\{r \in [0,1] | x \in M_r\} & \text{if } x \in M_0 \end{cases},$$

we obtain a fuzzy number (i.e., $u$ is a normal, fuzzy convex, upper semicontinuous and compactly supported fuzzy set). The level sets of $u$ are $u_r = M_r$, $r \in (0, 1]$ and $u_0 \subseteq M_0$. Now we prove these statements step by step.

"Normal:" Since $M_1$ is nonempty, for $x \in M_1$ we have $u(x_0) = 1$, for some $x_0 \in M_0$, so $u$ is normal.

"Fuzzy Convex:" In order to prove fuzzy convexity we consider now a fixed element in a fixed interval $x \in [a, b] \subseteq M_0$. Suppose that

$$u(a) = r_a = \sup\{r | a \in M_r\}$$

and

$$u(b) = r_b = \sup\{r | b \in M_r\}.$$

Suppose that $r_a \leq r_b$. Then from (ii) we have $M_{r_b} \subseteq M_{r_a}$ so $b \in M_{r_b} \subseteq M_{r_a}$. Now since both $a, b \in M_{r_a}$ and since $M_{r_a}$ are closed intervals it follows that $[a, b] \subseteq M_{r_a}$. Then

$$u(x) \geq r_a = u(a) = \min\{u(a), u(b)\}.$$

A symmetric reasoning can be followed in the case when $r_a \geq r_b$.

"$u_r = M_r$" Now let us prove that $u_r = \{x | u(x) \geq r\} = M_r$ by double inclusion. Let $r_0 \in (0, 1]$ be fixed and $x \in M_{r_0}$. Then $r_0 \in \{r | x \in M_r\}$. Then

$$u(x) = \sup\{r | x \in M_r\} \geq r_0$$

and this implies $x \in u_{r_0}$. So, we have obtained $M_{r_0} \subseteq u_{r_0}$. For the symmetric inclusion we consider $x \in u_{r_0}$, i.e., $u(x) \geq r_0$. Now let us suppose that strict inequality $u(x) > r_0$ holds. Then $\sup\{r | x \in M_r\} > r_0$ and there exists $r_1 \geq r_0$ with $x \in M_{r_1}$. Since $M_{r_1} \subseteq M_{r_0}$ according to $(ii)$, we obtain $x \in M_{r_0}$ which completes the reasoning in this case.

If we suppose

$$u(x) = r_0 = \sup\{r | x \in M_r\}$$

then there exists a sequence $r_n$ that converges from below to $r_0$ such that $x \in M_{r_n}, n \geq 1$. From (iii) we have

$$x \in \bigcap_{n=1}^{\infty} M_{r_n} = M_{r_0}.$$

As a conclusion we obtain $u_{r_0} \subseteq M_{r_0}$. So, $u_{r_0} = M_{r_0}$.

"Upper semicontinuity:" Since $u_r = M_r$ are closed according to (i) we have the complement of $u_r$, that is $\mathbb{R} \setminus u_r = \{x | u(x) < r\}$ an open set. This implies that $u$ is upper semicontinuous.

"Compact support:" It is easy to observe that

$$u_0 = cl\{x | u(x) > 0\} = cl \bigcup_{n=1}^{\infty} \{x | u(x) \geq r_n\} = cl \bigcup_{n=1}^{\infty} M_{r_n}$$

is closed and we have $u_0 = M_0$. Finally we obtain that $u_0$ is a bounded subset of the real line. Then, it is compact and the proof is complete. ∎

The following pair of Theorems is due to Goetschel and Voxman [74] and it gives another representation of a fuzzy number as a pair of functions that satisfy some properties. The representation provided by this theorem is called the LU (lower-upper) representation of a fuzzy number.

**Theorem 4.9.** *(LU-representation, Goetschel-Voxman [74]) Let $u$ be a fuzzy number and let $u_r = [u_r^-, u_r^+] = \{x | u(x) \geq r\}$. Then the functions $u^-, u^+$ : $[0,1] \to \mathbb{R}$, defining the endpoints of the $r-$level sets, satisfy the following conditions:*

*(i) $u^-(r) = u_r^- \in \mathbb{R}$ is a bounded, non-decreasing, left-continuous function in $(0,1]$ and it is right-continuous at 0.*

*(ii) $u^+(r) = u_r^+ \in \mathbb{R}$ is a bounded, non-increasing, left-continuous function in $(0,1]$ and it is right-continuous at 0.*

*(iii) $u_1^- \leq u_1^+$.*

**Proof.** For a given $u \in \mathbb{R}_{\mathcal{F}}$, and given $0 \leq r_1 \leq r_2 \leq 1$, from the stacking theorem we obtain $u_{r_2} \subseteq u_{r_1}$. Then we have

$$u_{r_1}^- \leq u_{r_2}^- \leq u_1^- \leq u_1^+ \leq u_{r_2}^+ \leq u_{r_1}^+,$$

$\forall r_1, r_2, 0 \leq r_1 \leq r_2 \leq 1$ which implies immediately the monotonicity properties and (iii). Left continuity at $r \in (0,1]$ follows from property (iii) of the Stacking theorem. Indeed, let $r_0 \in (0,1]$ be fixed and $r_n$ converging from below to $r_0$, i.e., $r_n \uparrow r_0$. Then from the property (iii) of the Stacking Theorem we obtain

$$\bigcap_{n=1}^{\infty} u_{r_n} = u_{r_0},$$

which immediately implies $u_{r_n}^- \to u_{r_0}^-$ and $u_{r_n}^+ \to u_{r_0}^+$, i.e., both functions are left continuous at arbitrary $r_0 \in (0,1]$. In order to prove right continuity at 0 we consider $r_n \downarrow 0$, a sequence that converges from above to 0. We have

$$u_0 = cl\{x | u(x) > 0\} = cl \bigcup_{n=1}^{\infty} \{x | u(x) \geq r_n\} = cl \left( \bigcup_{n=1}^{\infty} u_{r_n} \right).$$

Since $u_0$ is compact then it is closed and bounded and we get $u_{r_n}^- \to u_0^-$ and $u_{r_n}^+ \to u_0^+$, which implies right continuity at 0. ∎

The reciprocal of the LU-representation is the Goetschel-Voxman characterization theorem.

**Theorem 4.10.** *(Goetschel-Voxman [74]) Let us consider the functions $u^-, u^+$ : $[0,1] \to \mathbb{R}$, that satisfy the following conditions:*

*(i) $u^-(r) = u_r^- \in \mathbb{R}$ is a bounded, non-decreasing, left-continuous function in $(0,1]$ and it is right-continuous at 0.*

*(ii)* $u^+(r) = u_r^+ \in \mathbb{R}$ *is a bounded, non-increasing, left-continuous function in* $(0,1]$ *and it is right-continuous at* 0.

*(iii)* $u_1^- \leq u_1^+$.

Then there is a fuzzy number $u \in \mathbb{R}_{\mathcal{F}}$ that has $u_r^-, u_r^+$ as endpoints of its $r$−level sets, $u_r$.

**Proof.** We will prove that the sets $M_r = [u_r^-, u_r^+]$ satisfy the conditions in Negoita-Ralescu characterization theorem and then this family of subsets defines a fuzzy number $u \in \mathbb{R}_{\mathcal{F}}$. From the monotonicity properties and from (iii) we immediately obtain $u_{r_1}^- \leq u_{r_2}^- \leq u_1^- \leq u_1^+ \leq u_{r_2}^+ \leq u_{r_1}^+$ and this implies $M_r = [u_r^-, u_r^+]$ are non-empty, closed intervals and that $M_{r_2} \subseteq M_{r_1}$, for any $r_1 \leq r_2$. Let us consider now a sequence $r_n$ which converges from below to $r \in (0,1]$. Then from the left continuity properties we obtain $u_{r_n}^- \to u_r^-$ and $u_{r_n}^+ \to u_r^+$ and then we have

$$\bigcap_{n=1}^{\infty} M_{r_n} = M_r.$$

From the right continuity at 0 we obtain

$$cl\{x \in \mathbb{R} : u(x) > 0\} = M_0.$$

Then there exists $u \in \mathbb{R}_{\mathcal{F}}$, such that

$$u_r = M_r = [u_r^-, u_r^+],$$

for any $r \in [0,1]$. ∎

**Remark 4.11.** *The conditions in the definition of fuzzy numbers are shown to be too restrictive in some applications. There are several new research directions that are generalizing, revising the fuzzy number concept in different directions (see Fortin-Dubois-Fargier [62], Bica [29], etc). This represents an interesting topic for further investigation.*

## 4.3 L-R Fuzzy Numbers

The so-called *L-R* fuzzy numbers are considered important in the theory of fuzzy sets. L-R fuzzy numbers, and their particular cases, as e.g., triangular and trapezoidal fuzzy numbers, are very useful in applications.

**Definition 4.12.** *(Dubois-Prade [50]) Let* $L, R : [0,1] \to [0,1]$ *be two continuous, increasing functions fulfilling* $L(0) = R(0) = 0, L(1) = R(1) = 1$.

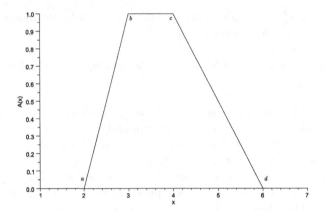

**Fig. 4.6** Example of a trapezoidal fuzzy number

Let $a_0^- \leq a_1^- \leq a_1^+ \leq a_0^+$ be real numbers. The fuzzy set $u : \mathbb{R} \to [0,1]$ is an *L-R* **fuzzy number** if

$$
u(x) = \begin{cases}
0 & \text{if} & x < a_0^- \\
L\left(\frac{x - a_0^-}{a_1^- - a_0^-}\right) & \text{if} & a_0^- \leq x < a_1^- \\
1 & \text{if} & a_1^- \leq x < a_1^+ \\
R\left(\frac{a_0^+ - x}{a_0^+ - a_1^+}\right) & \text{if} & a_1^+ \leq x < a_0^+ \\
0 & \text{if} & a_0^+ \leq x
\end{cases}.
$$

Symbolically, we write $u = \left(a_0^-, a_1^-, a_1^+, a_0^+\right)_{L,R}$, where $[a_1^-, a_1^+]$ is the core of $u$, and $\underline{a} = a_1^- - a_0^-, \bar{a} = a_0^+ - a_1^+$ are called the left and the right spread respectively. If $u$ is an *L-R* fuzzy number then its level sets can be calculated as

$$
u_r = \left[a_0^- + L^{-1}(r)\,\underline{a}, \, a_0^+ - R^{-1}(r)\,\bar{a}\right].
$$

As a particular case, one obtains **trapezoidal fuzzy numbers** when the functions $L$ and $R$ are linear. A trapezoidal fuzzy number $u$ can be represented by the quadruple $(a, b, c, d) \in \mathbb{R}^4$, $a \leq b \leq c \leq d$, (see Fig. 4.6)

$$
u(t) = \begin{cases}
0 & \text{if} & t < a \\
\frac{t - a}{b - a} & \text{if} & a \leq t < b \\
1 & \text{if} & b \leq t \leq c \\
\frac{d - t}{d - c} & \text{if} & c < t \leq d \\
0 & \text{if} & d < t
\end{cases}.
$$

In this case the endpoints of the $r$-level sets are given by

$$
u_r^- = a + r(b - a)
$$

and

$$
u_r^+ = d - r(d - c).
$$

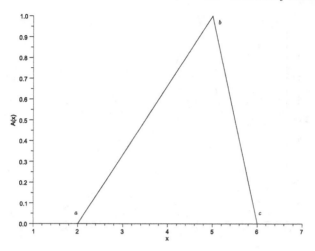

**Fig. 4.7** Example of a triangular fuzzy number

If we have $b = c$ in the representation $(a, b, c, d)$, the fuzzy number is called a **triangular fuzzy number**. Then a triplet $(a, b, c) \in \mathbb{R}^3$, $a \le b \le c$ is sufficient to represent the fuzzy number (see Figure 4.7).

**Example 4.13.** *Let us consider* $L, R : [0, 1] \to [0, 1]$, $L(x) = R(x) = x^2$. *Let* $a_1^- = a_1^+ = 1$ *and* $\underline{a} = 1, \bar{a} = 2$ *respectively. Then the* $L - R$ *fuzzy number determined by Definition 4.12 has the level sets*

$$u_r = \left[ \sqrt{r}, 3 - 2\sqrt{r} \right]$$

*and the shape depicted in Fig. 4.8.*

We will consider in what follows other useful classes of fuzzy numbers.

A **Gaussian fuzzy number** has the membership degree

$$
u(x) = \begin{cases}
0 & \text{if} \quad x < x_1 - a\sigma_l \\
e^{-\frac{(x - x_1)^2}{2\sigma_l^2}} & \text{if} \quad x_1 - a\sigma_l \le x < x_1 \\
e^{-\frac{(x_1 - x)^2}{2\sigma_r^2}} & \text{if} \quad x_1 \le x < x_1 + a\sigma_r \\
0 & \text{if} \quad x_1 + a\sigma_r \le x
\end{cases},
$$

where $x_1$ is the core of the fuzzy number, $\sigma_l, \sigma_r$ are the left and right spreads and $a > 0$ is a tolerance value. Its graph is shown in Figure 4.9. Often in fuzzy control applications we have $a \to \infty$ in which case the fuzzy set is not a fuzzy number since it fails to have a compact support, however it is very useful in fuzzy control systems.

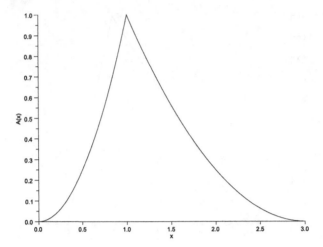

**Fig. 4.8** Example of an $L - R$ fuzzy number

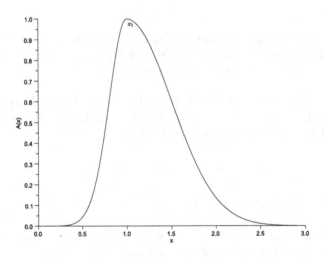

**Fig. 4.9** Example of a Gaussian Fuzzy number

An **exponential fuzzy number** has the membership degree given by

$$
u(x) = \begin{cases}
0 & \text{if} & x < x_1 - a\tau_l \\
e^{\frac{x - x_1}{\tau_l}} & \text{if} & x_1 - a\tau_l \leq x < x_1 \\
e^{\frac{x_1 - x}{\tau_r}} & \text{if} & x_1 \leq x < x_1 + a\tau_r \\
0 & \text{if} & x_1 + a\tau_r \leq x
\end{cases},
$$

where $x_1$ is the core of the exponential fuzzy number $\tau_l$, $\tau_r$ are the left and right spread of it respectively, and $a$ represents a tolerance value. Its graph is shown in Figure 4.10.

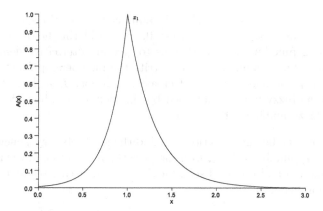

**Fig. 4.10** Example of an exponential fuzzy number

**Remark 4.14.** *In some applications, to have a compact support is not important. Then we will consider fuzzy sets that are not fuzzy numbers in the proper sense of the term, but they are very simple and useful in applications. Moreover, if e.g. the Universe of discourse is restricted to a closed compact interval, then all the fuzzy sets that fulfill (i), (ii) and (iii) of Definition 4.1 become fuzzy numbers. As examples of such fuzzy sets we mention fuzzy sets with Gaussian membership function*

$$u(x) = e^{-\frac{(x-x_1)^2}{2\sigma^2}}$$

*and fuzzy sets with exponential membership function*

$$u(x) = e^{-\frac{|x-x_1|}{\tau}}.$$

**Remark 4.15.** *In this section we investigated $L-R$ fuzzy numbers, which were based on Negoita-Ralescu characterization. Similarly $L-U$ representation of fuzzy sets was recently investigated in several papers as e.g., Stefanini-Sorini-Guerra[137], Stefanini[135]. Another interesting idea within the topic of the present chapter is the problem of approximation of a fuzzy number by trapezoidal or other types of fuzzy numbers (Grzegorzewski-Mrówka [75], Ban-Coroianu-Grzegorzewski[11]).*

## 4.4   Problems

1. Prove that if $u$ is an L-R fuzzy number then its level sets are given by

$$u_r = \left[ a_0^- + L^{-1}(r)\underline{a}, a_0^+ - R^{-1}(r)\overline{a} \right], r \in [0,1].$$

2. In the definition of an $L - R$ fuzzy number the functions $L$ and $R$ are considered increasing $L, R : [0, 1] \to [0, 1]$. Considering however $L, R$ two decreasing functions would allow us to give an alternative definition of the $L - R$ fuzzy number concept. Write the membership function of an $L - R$ fuzzy number based on decreasing functions $L, R$. Find the level-sets of the fuzzy number obtained by this definition. Show examples of $L - R$ fuzzy numbers of this type.

3. Write the membership function and find the level sets for a general triangular fuzzy number $u = (a, b, c)$, then use the results to find the membership function and the level sets, for the particular case of the triangular fuzzy number $(2, 4, 5)$.

4. Write the membership function for the L-R fuzzy number in Example 4.13 and prove that its level sets are given by

$$u_r = \left[ \sqrt{r}, 3 - 2\sqrt{r} \right].$$

5. Given the sets $M_r = \left[ e + e^r, 4e - e^{r^2} \right]$ show that these sets define a fuzzy number such that $u_r = M_r$. Find the expression of the membership function of $u$.

6. Set up triangular, Gaussian and exponential membership functions for fuzzy sets that model the linguistic expression "about 100".

7. Find an expression for the level sets of a Gaussian fuzzy number.

8. Find an expression for the level sets of an exponential fuzzy number.

9. Consider $L(x) = R(x) = x^a, a > 0$. Write the membership function and find the level sets of the $L - R$ fuzzy number based on these functions.

10. Find a function of the form $a + b \cos(cx + d)$ such that $L, R$ will be correct shape functions for an $L - R$ fuzzy number and additionally, such that the $L - R$ fuzzy numbers that we obtain will have smooth membership functions. Write the membership function and the level sets of the $L - R$ fuzzy number obtained.

# 5
# Fuzzy Arithmetic

## 5.1  Zadeh's Extension Principle

Often, we have to perform operations with uncertain parameters. In this case we will have to define the fuzzy counterparts of the classical operations between real numbers.

We begin our discussion on fuzzy arithmetic with Zadeh's extension principle. It serves for extending a real-valued function into a corresponding fuzzy function.

**Definition 5.1.** *(Zadeh's Extension Principle, Zadeh [156]) Given a function $f : X \to Y$, where $X$ and $Y$ are crisp sets, it can be extended to a function $F : \mathcal{F}(X) \to \mathcal{F}(Y)$ (a fuzzy function) such that $v = F(u)$, where*

$$v(y) = \begin{cases} \sup\{u(x) : x \in X, f(x) = y\}, & \text{when } f^{-1}(y) \neq \varnothing \\ 0, & \text{otherwise} \end{cases}.$$

*We call the function $F$ as **Zadeh's extension** of $f$.*

**Remark 5.2.** *Zadeh's extension is well defined for any fuzzy set $u \in \mathcal{F}(X)$. Indeed, when $f^{-1}(y) \neq \varnothing$, the set $\{u(x) : x \in X, f(x) = y\}$ is non-empty and bounded and so, it has a least upper bound.*

Naturally raises the question whether a fuzzy number in the input generates a fuzzy number in the output.

**Theorem 5.3.** *(Nguyen [116], Barros-Bassanezi-Tonelli [12], Fullér-Keresztfalvi [65]) Given a continuous $f : \mathbb{R} \to \mathbb{R}$, it can be extended to a fuzzy*

B. Bede: *Mathematics of Fuzzy Sets and Fuzzy Logic*, STUDFUZZ 295, pp. 65–78.
DOI: 10.1007/978-3-642-35221-8_5      © Springer-Verlag Berlin Heidelberg 2013

*function $F : \mathbb{R}_{\mathcal{F}} \to \mathbb{R}_{\mathcal{F}}$, and given $u \in \mathbb{R}_{\mathcal{F}}$ we can determine $v = F(u) \in \mathbb{R}_{\mathcal{F}}$ by its level sets $v_r = F(u_r)$, $\forall r \in [0,1]$, i.e., we have $v_r = [v_r^-, v_r^+]$, where*

$$v_r^- = \inf\{f(x)|x \in u_r\}$$
$$v_r^+ = \sup\{f(x)|x \in u_r\}.$$

$u_r, r \in [0,1]$ *denote the level sets of $u$.*

**Proof.** First let us prove that if $u_r$ and $v_r$ are level sets of the fuzzy sets $u$ and $v = F(u)$ respectively then $v_r = f(u_r)$.

The case $f^{-1}(y) = \varnothing$ is obvious. If $f^{-1}(y) \neq \varnothing$ we have $v = F(u)$ given as

$$v(y) = \sup\{u(x)|x \in X, f(x) = y\}.$$

If $x \in u_r$ then $u(x) \geq r$ and it implies also that $v(y) \geq r$ and so $y = f(x) \in v_r$, i.e., $f(u_r) \subseteq v_r$. On the other hand if $v(y) \geq r$ then for any $\varepsilon > 0$ there exists an $x \in f^{-1}(y)$ such that
$$v(y) - \varepsilon < u(x)$$

which implies $u(x) \geq r$ and so, we obtain that Zadeh's extension satisfies $v_r \subseteq f(u_r)$ so finally $v_r = f(u_r)$. So, if $v_r = [v_r^-, v_r^+]$, we obtain

$$v_r^- = \inf\{f(x)|x \in u_r\}$$
$$v_r^+ = \sup\{f(x)|x \in u_r\}.$$

Let us prove now that if $u_r$ are level sets of a fuzzy number $u \in \mathbb{R}_{\mathcal{F}}$ then $v_r = f(u_r)$ define level sets of a fuzzy number $v \in \mathbb{R}_{\mathcal{F}}$, $v = F(u)$. For this aim we will prove that $v_r$ satisfy the hypotheses of the Negoita-Ralescu characterization. First we observe that since $u_r$ are compact convex intervals in $\mathbb{R}$ and since $f$ is continuous, we get $v_r = f(u_r)$ compact convex. This means that $v_r$ is a closed interval for any $r \in [0,1]$.

If $r \leq s$ then we have $u_s \subseteq u_r$. This implies

$$v_s = f(u_s) \subseteq f(u_r) = v_r.$$

Let us consider the sequence $r_n$ that converges from below to $r$. Then

$$\bigcap_{n=1}^{\infty} u_{r_n} = u_r.$$

We will prove that

$$\bigcap_{n=1}^{\infty} v_{r_n} = v_r,$$

i.e.,

$$\bigcap_{n=1}^{\infty} f(u_{r_n}) = f(u_r)$$

which is equivalent to

$$\bigcap_{n=1}^{\infty} f(u_{r_n}) = f\left(\bigcap_{n=1}^{\infty} u_{r_n}\right).$$

Let

$$y \in f\left(\bigcap_{n=1}^{\infty} u_{r_n}\right).$$

Then there exist an $x \in u_{r_n}, \forall n = 1, 2, \ldots$ such that $y = f(x)$. Then $y \in f(u_{r_n}), \forall n = 1, 2, \ldots$, i.e.,

$$y \in \bigcap_{n=1}^{\infty} f(u_{r_n}).$$

On the other hand let

$$y \in \bigcap_{n=1}^{\infty} f(u_{r_n}).$$

Then $y \in f(u_{r_n}), \forall n = 1, 2, \ldots$ Then there exists an $x_n \in u_{r_n}$ with $y = f(x_n), \forall n = 1, 2, \ldots$ Since $x_n$ is a bounded sequence and since $x_n \in u_0$ with $u_0$ compact, we get the existence of a sub-sequence $x_{n_k}, k = 1, 2, \ldots$ that converges. Let $x = \lim_{k \to \infty} x_{n_k}$. It is easy to see that $x \in u_r$. Indeed, $u_r \subseteq u_{r_n}, \forall n = 1, 2, \ldots$ and if we suppose that $x \notin u_r$ we get that there exists $k$ such that $x \notin u_k$. Taking into account that $x = \lim_{k \to \infty} x_{n_k}$ we obtain a contradiction.

Combining the two inclusions we obtain

$$\bigcap_{n=1}^{\infty} v_{r_n} = v_r.$$

Let us consider the sequence $r_n$ that converges from above to 0. Then it is known that

$$cl\left(\bigcup_{n=1}^{\infty} u_{r_n}\right) = u_0.$$

It is easy to check that since $f$ is continuous we have $f(cl(A)) \subseteq cl(f(A))$, and then

$$v_0 = f\left(cl\left(\bigcup_{n=1}^{\infty} u_{r_n}\right)\right) \subseteq cl\left(\bigcup_{n=1}^{\infty} f(u_{r_n})\right) = cl\left(\bigcup_{n=1}^{\infty} v_{r_n}\right),$$

which immediately implies

$$v_0 = cl\left(\bigcup_{n=1}^{\infty} v_{r_n}\right),$$

and so, the hypotheses of Negoita-Ralescu characterization theorem are fulfilled. Using the theorem we have proved that $v_r$ are level sets of a fuzzy number, i.e., $v \in \mathbb{R}_{\mathcal{F}}$.    ∎

The two dimensional case of Zadeh's extension principle is very important because it allows us to extend operations between real numbers to the fuzzy case.

**Definition 5.4.** *In the case of two variables ($f : X \times Y \to Z$ ) the function can be extended to $F : \mathcal{F}(X) \times \mathcal{F}(Y) \to \mathcal{F}(Z)$ such that $w = F(u, v)$, where*

$$w(z) = \begin{cases} \sup_{x \in X, y \in Y} \{\min\{u(x), v(y)\} f(x, y) = z\} \text{ if } f^{-1}(z) \neq \varnothing \\ 0, \text{ otherwise} \end{cases}.$$

**Remark 5.5.** *We should exercise however caution, because there are pitfalls as it was shown in Huang [78]. Namely, an extension principle based on compact convex subsets of $\mathbb{R}^2$ is not possible. The following correct extension result allows us to extend real operations to the fuzzy number's case.*

**Theorem 5.6.** *(Nguyen [116], Barros-Bassanezi-Tonelli [12], Fullér-Keresztfalvi [65]) If we assume that $f : \mathbb{R} \times \mathbb{R} \to \mathbb{R}$, continuous, then we can extend it to $F : \mathbb{R}_{\mathcal{F}} \times \mathbb{R}_{\mathcal{F}} \to \mathbb{R}_F$ such that $w = F(u, v)$ has its level sets*

$$w_r = \{f(x, y) | x \in u_r, y \in v_r\}.$$

*for any $u, v \in \mathbb{R}_{\mathcal{F}}$, i.e. if $w_r = [w_r^-, w_r^+]$ then*

$$w_r^- = \inf\{f(x, y) | x \in u_r, y \in v_r\}$$

$$w_r^+ = \sup\{f(x, y) | x \in u_r, y \in v_r\}.$$

**Proof.** First let us prove that if $u_r, v_r$ and $w_r$ are level sets of the fuzzy sets $u, v$ and $w = F(u, v)$ respectively then $w_r = f(u_r, v_r)$.

The case $f^{-1}(z) = \varnothing$ is obvious. If $f^{-1}(z) \neq \varnothing$ we have $w = F(u, v)$ given as

$$w(y) = \sup\{\min\{u(x), v(y)\} : x \in X, y \in Y, f(x, y) = z\}.$$

If $x \in u_r$, $y \in v_r$ then $u(x) \geq r$, $v(y) \geq r$ and for $z = f(x, y)$ it implies also that $w(z) \geq r$ i.e., $f(u_r, v_r) \subseteq w_r$. On the other hand if $w(y) \geq r$ then for any $\varepsilon > 0$ there exists an $(x, y) \in f^{-1}(z)$ such that

$$w(z) - \varepsilon < u(x) \text{ and } w(z) - \varepsilon < v(y)$$

which implies $u(x) \geq r$ and $v(y) \geq r$ so, we obtain $w_r = f(u_r, v_r)$. From now on the reasoning is similar to that of the preceding result.

Since $u_r, v_r$ are compact convex intervals in $\mathbb{R}$ and since $f$ is continuous, we get $w_r = f(u_r, v_r)$ compact convex.

If $r \leq s$ then we have $u_s \subseteq u_r$ and $v_s \subseteq v_r$. This implies

$$w_s = f(u_s, v_s) \subseteq f(u_r, v_r) = w_r.$$

Let now $r_n$ be a sequence that converges from below to $r$. Then the relations

$$\bigcap_{n=1}^{\infty} w_{r_n} = w_r$$

can be obtained in a similar way as in the previous theorem. Finally, if $r_n$ converges from above to 0 we get

$$w_0 = cl \left( \bigcup_{n=1}^{\infty} w_{r_n} \right),$$

and so, the hypotheses of Negoita-Ralescu characterization theorem are fulfilled and finally we obtain $w \in \mathbb{R}_\mathcal{F}$. ∎

## 5.2 The Sum and Scalar Multiplication

For $u, v \in \mathbb{R}_\mathcal{F}$ and $\lambda \in \mathbb{R}$, based on the extension principle, one can define the sum of two fuzzy numbers $u + v$ and the multiplication between a real and a fuzzy number $\lambda \cdot u$. Then by Theorem 5.6, since the sum and scalar multiplication are continuous functions, we obtain,

$$(u + v)_r = \{x + y | x \in u_r, y \in v_r\} = u_r + v_r,$$

$$(\lambda \cdot u)_r = \{\lambda x | x \in u_r\} = \lambda u_r, \forall r \in [0, 1],$$

where $u_r + v_r$ is the sum of two intervals (as subsets of $\mathbb{R}$), and $\lambda u_r$ is the usual product of a number and a subset of $\mathbb{R}$. So, fuzzy arithmetic extends interval arithmetic.

**Example 5.7.** Let $u = (1, 2, 3)$, $v = (2, 3, 5)$ be triangular fuzzy numbers then $u + v = (3, 5, 8)$. Also we have $2u = (2, 4, 6)$ and $-2v = (-10, -6, -4)$.

The following Theorem deals with the algebraic properties of $\mathbb{R}_\mathcal{F}$.

**Theorem 5.8.** (Anastassiou-Gal [5], Dubois-Prade [50], Gal [69]) (i) The addition of fuzzy numbers is associative and commutative i.e.,

$$u + v = v + u$$

and

$$u + (v + w) = (u + v) + w,$$

$\forall u, v, w \in \mathbb{R}_\mathcal{F}$.

*(ii) The singleton fuzzy set* $0 = \mathcal{X}_{\{0\}} \in \mathbb{R}_{\mathcal{F}}$ *is the neutral element w.r.t.* $+$ *i.e.,*

$$u + 0 = 0 + u = u,$$

*for any* $u \in \mathbb{R}_{\mathcal{F}}$.

*(iii) None of* $u \in \mathbb{R}_{\mathcal{F}} \setminus \mathbb{R}$ *has an opposite in* $\mathbb{R}_{\mathcal{F}}$ *(w.r.t.* $+$*).*

*(iv) For any* $a, b \in \mathbb{R}$ *with* $a \cdot b \geq 0$ *and any* $u \in \mathbb{R}_{\mathcal{F}}$, *we have*

$$(a + b) \cdot u = (a \cdot u) + (b \cdot u) .$$

*For general* $a, b \in \mathbb{R}$, *this property does not hold.*

*(v) For any* $\lambda \in \mathbb{R}$ *and* $u, v \in \mathbb{R}_{\mathcal{F}}$, *we have*

$$\lambda \cdot (u + v) = \lambda \cdot u + \lambda \cdot v.$$

*(vi) For any* $\lambda, \mu \in \mathbb{R}$ *and any* $u \in \mathbb{R}_{\mathcal{F}}$, *we have*

$$(\lambda \cdot \mu) \cdot u = \lambda \cdot (\mu \cdot u)$$

**Proof.** The properties are easy to be verified. Also, it is easy to see that $(2-1)(1, 2, 3) = (1, 2, 3)$, while $2(1, 2, 3) - (1, 2, 3) = (-1, 2, 5)$, where $u - v = u + (-1)v$. ∎

As a conclusion from the previous Theorem we obtain that the space of fuzzy numbers is not a linear space.

## 5.3   The Product of Two Fuzzy Numbers

The product $w = u \cdot v$ of fuzzy numbers $u$ and $v$, is defined (see e.g. Hanss [77]) based on Zadeh's extension principle. Using again Theorem 5.6 we have it defined by its endpoints as

$$w_r^- = \inf\{x \cdot y | x \in u_r, y \in v_r\}$$

$$w_r^+ = \sup\{x \cdot y | x \in u_r, y \in v_r\}.$$

The product attains its extrema at the corners of its domain. Then,

$$(u \cdot v)_r^- = \min\{u_r^- v_r^-, u_r^- v_r^+, u_r^+ v_r^-, u_r^+ v_r^+\}$$

and

$$(u \cdot v)_r^+ = \max\{u_r^- v_r^-, u_r^- v_r^+, u_r^+ v_r^-, u_r^+ v_r^+\}.$$

**Example 5.9.** *As an example, consider* $u = (0, 2, 4, 6)$ *and* $v = (2, 3, 8)$. *Then the endpoints if the level sets of* $u \cdot v$ *are*

$$(u \cdot v)_r^- = 2r \cdot (r + 2)$$

$$(u \cdot v)_r^+ = (6 - 2r) \cdot (8 - 5r)$$

*which leads* $(u \cdot v)_1 = [6, 12]$, $(u \cdot v)_0 = [0, 48]$, *and membership function shown in Figure 5.1.*

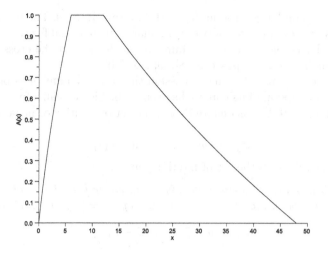

**Fig. 5.1** Product of two fuzzy numbers

The following Theorem discusses the algebraic properties of $\mathbb{R}_{\mathcal{F}}$ related to the multiplication.

**Theorem 5.10.** *(i) The singleton fuzzy set* $1 = \chi_{\{1\}} \in \mathbb{R}_{\mathcal{F}}$ *is the neutral element w.r.t.* $\cdot$ *i.e.,* $u \cdot 1 = 1 \cdot u = u$, *for any* $u \in \mathbb{R}_{\mathcal{F}}$.
*(ii) None of* $u \in \mathbb{R}_{\mathcal{F}} \setminus \mathbb{R}$ *has an inverse in* $\mathbb{R}_{\mathcal{F}}$ *(w.r.t.* $\cdot$ *).*
*(iii) For any* $u, v, w \in \mathbb{R}_{\mathcal{F}}$ *we have*

$$((u + v) \cdot w)_r \subseteq (u \cdot w)_r + (v \cdot w)_r, \forall r \in [0, 1],$$

*and, in general, distributivity does not hold.*
*(v) For any* $u, v, w \in \mathbb{R}_{\mathcal{F}}$ *are such that none of the supports of* $u, v, w$ *does not contain* 0, *we have*

$$u \cdot (v \cdot w) = (u \cdot v) \cdot w .$$

**Proof.** The properties are easily verified. Let us prove (iii). From Theorem 5.6 we have

$$((u + v) \cdot w)_r^- = \min\{(u_r^- + v_r^-) \cdot w_r^-, (u_r^- + v_r^-)$$

$$\cdot w_r^+, (u_r^+ + v_r^+) \cdot w_r^-, (u_r^+ + v_r^+) \cdot w_r^+\}$$

$$\geq \min\{u_r^- \cdot w_r^-, u_r^- \cdot w_r^+, u_r^+ \cdot w_r^-, u_r^+ \cdot w_r^+\}$$

$$+ \min\{v_r^- \cdot w_r^-, v_r^- \cdot w_r^+, v_r^+ \cdot w_r^-, v_r^+ \cdot w_r^+\},$$

which combined with the symmetric inequality leads to

$$((u + v) \cdot w)_r \subseteq (u \cdot w)_r + (v \cdot w)_r, \forall r \in [0, 1].$$

∎

Zadeh's extension based product has certain disadvantages. The most important one is that the product of two triangular or trapezoidal fuzzy numbers is not triangular trapezoidal, etc. To improve on this aspect the cross product of fuzzy numbers was proposed in [8], [9] and [10].

We will say in this following discussion that a fuzzy number is positive if for the lower endpoint of its core we have $w_1^- \geq 0$. Also we call a fuzzy number negative if $w_1^+ \leq 0$. Let us denote by $\mathbb{R}_{\mathcal{F}}^*$ the set of positive or negative fuzzy numbers, i.e.,

$$\mathbb{R}_{\mathcal{F}}^* = \{w \in \mathbb{R}_{\mathcal{F}} | 0 \notin \text{int}(w_1)\},$$

where $\text{int}(w_1)$ denotes the set of interior points of $w_1$.

**Proposition 5.11.** *(Ban-Bede [8], [9], Bede-Fodor [17]) If $u$ and $v$ are positive fuzzy numbers, then $w = u \odot v$ defined by $w_r = [w_r^-, w_r^+]$, where*

$$w_r^- = u_r^- v_1^- + u_1^- v_r^- - u_1^- v_1^-$$

*and*

$$w_r^+ = u_r^+ v_1^+ + u_1^+ v_r^+ - u_1^+ v_1^+,$$

*for every $r \in [0, 1]$, is a positive fuzzy number.*

**Proof.** We have

$$w_r^+ - w_r^- = \left(u_r^+ - u_1^+\right) v_1^+ + \left(u_1^- - u_r^-\right) v_1^-$$

$$+ u_1^+ v_r^+ - u_1^- v_r^- \geq 0,$$

for every $r \geq 0$, and so $w_r = [w_r^-, w_r^+]$, is a closed interval. This verifies (i) in Theorem 4.8.

Let us consider $r_1, r_2 \in [0, 1]$, $r_1 \leq r_2$. Because $u_{r_2} \subseteq u_{r_1}$ and $v_{r_2} \subseteq v_{r_1}$ we obtain

$$w_{r_1}^- = u_{r_1}^- v_1^- + u_1^- v_{r_1}^- - u_1^- v_1^-$$

$$\leq u_{r_2}^- v_1^- + u_1^- v_{r_2}^- - u_1^- v_1^- = w_{r_2}^-$$

and

$$w_{r_1}^+ = u_{r_1}^+ v_1^+ + u_1^+ v_{r_1}^+ - u_1^+ v_1^+$$

$$\leq u_{r_2}^+ v_1^+ + u_1^+ v_{r_2}^+ - u_1^+ v_1^+ = w_{r_2}^+$$

which implies (ii) of Theorem 4.8.

Let us consider $(r_n)_{n \in \mathbf{N}}$ converging increasingly to $r \in [0, 1]$. The conditions in Theorem 4.8 imply

$$u_{r_n}^- v_1^- + u_1^- v_{r_n}^- - u_1^- v_1^- \searrow u_r^- v_1^- + u_1^- v_r^- - u_1^- v_1^-$$

and

$$u_{r_n}^+ v_1^+ + u_1^+ v_{r_n}^+ - u_1^+ v_1^+ \nearrow u_r^+ v_1^+ + u_1^+ v_r^+ - u_1^+ v_1^+.$$

such that we obtain (iii) of Theorem 4.8. The proof of (iv) is similar to (iii). The conclusion of the Theorem 4.8 is that the level sets $w_r$ define a fuzzy number. ∎

**Proposition 5.12.** *Let $u$ and $v$ two fuzzy numbers.*
*(i) If $u$ is positive and $v$ is negative then*

$$u \odot v = -\left(u \odot (-v)\right)$$

*is a negative fuzzy number.*
*(ii) If $u$ is negative and $v$ is positive then*

$$u \odot v = -\left((-u) \odot v\right)$$

*is a negative fuzzy number.*
*(iii) If $u$ and $v$ are negative then*

$$u \odot v = (-u) \odot (-v)$$

*is a positive fuzzy number.*

**Proof.** It is left to the reader as an exercise.    ∎

**Definition 5.13.** *(Ban-Bede [8]) The binary operation on $\mathbb{R}_{\mathcal{F}}^{*}$ introduced as above is called the cross product of fuzzy numbers.*

The cross product is defined for any fuzzy numbers in

$$\mathbb{R}_{\mathcal{F}}^{\wedge} = \{u \in \mathbb{R}_{\mathcal{F}}^{*}; \text{ there exists an unique } x_0 \in \mathbb{R} \text{ such that } u(x_0) = 1\},$$

therefore it is well defined for triangular fuzzy numbers.

**Example 5.14.** *The cross product of two positive triangular fuzzy numbers $u = (a_1, b_1, c_1)$ and $v = (a_2, b_2, c_2)$ is*

$$u \odot v = (a_1 b_2 + a_2 b_1 - b_1 b_2, b_1 b_2, b_1 c_2 + b_2 c_1 - b_1 c_1).$$

*For example if $u = (2, 3, 4)$ and $v = (3, 4, 6)$ is*

$$u \odot v = (5, 12, 22).$$

## 5.4   Difference of Fuzzy Numbers

The standard difference induced by Zadeh's extension principle, has the property $u - u \neq 0$. This is a shortcoming in some theoretical results and applications of fuzzy numbers. To avoid this shortcoming several new differences were proposed.

**Definition 5.15.** *(Hukuhara [79], Puri-Ralescu [123]) The Hukuhara difference (H-difference $\ominus_H$) is defined by*

$$u \ominus_H v = w \iff u = v + w,$$

*being $+$ the standard fuzzy addition.*

If $u \ominus_H v$ exists, its $\alpha - cuts$ are

$$[u \ominus_H v]_\alpha = [u_\alpha^- - v_\alpha^-, u_\alpha^+ - v_\alpha^+].$$

It is easy to verify that $u \ominus_H u = 0$ for any fuzzy numbers $u$, but as we have earlier discussed $u - u \neq 0$.

The Hukuhara difference rarely exists, so several alternatives and generalizations were proposed as e.g. the generalized Hukuhara differentiability.

**Definition 5.16.** *(Stefanini [136], Stefanini-Bede [141]) Given two fuzzy numbers $u, v \in \mathbb{R}_\mathcal{F}$, the generalized Hukuhara difference (gH-difference for short) is the fuzzy number $w$, if it exists, such that*

$$u \ominus_{gH} v = w \iff \left\{ \begin{array}{lll} & (i) & u = v + w \\ or & (ii) & v = u - w \end{array} \right. .$$

In terms of $\alpha$-cuts we have

**Proposition 5.17.** *For any $u, v \in \mathbb{R}_\mathcal{F}$ we have*

$$[u \ominus_{gH} v]_\alpha = [\min\{u_\alpha^- - v_\alpha^-, u_\alpha^+ - v_\alpha^+\}, \max\{u_\alpha^- - v_\alpha^-, u_\alpha^+ - v_\alpha^+\}].$$

**Proof.** The proof is left to the reader as an exercise.    ∎

The level-wise expression that we have obtained is the interval arithmetic proposed in Markov [108], [109].

The generalized Hukuhara difference exists in many more situations than the usual Hukuhara difference, but it does not always exist. We consider the following generalized difference (g-difference) that presents some advantages.

**Definition 5.18.** *(Stefanini [136], Bede-Stefanini [27])The generalized difference (g-difference for short) of two fuzzy numbers $u, v \in \mathbb{R}_\mathcal{F}$ is given by its level sets as*

$$[u \ominus_g v]_\alpha = cl \bigcup_{\beta \geq \alpha} ([u]_\beta \ominus_{gH} [v]_\beta), \forall \alpha \in [0, 1],$$

*where the gH-difference $\ominus_{gH}$ is with interval operands $[u]_\beta$ and $[v]_\beta$ and it is well defined in this case.*

**Proposition 5.19.** *(Bede-Stefanini [27]) The g-difference is given by the expression*

$$[u \ominus_g v]_\alpha = \left[ \inf_{\beta \geq \alpha} \min\{u_\beta^- - v_\beta^-, u_\beta^+ - v_\beta^+\}, \sup_{\beta \geq \alpha} \max\{u_\beta^- - v_\beta^-, u_\beta^+ - v_\beta^+\} \right] .$$

**Proof.** Let $\alpha \in [0,1]$ be fixed. We observe that for any $\beta \geq \alpha$ we have

$$[u]_\beta \ominus_{gH} [v]_\beta = \left[ \min\{u_\beta^- - v_\beta^-, u_\beta^+ - v_\beta^+\}, \max\{u_\beta^- - v_\beta^-, u_\beta^+ - v_\beta^+\} \right]$$

$$\subseteq \left[ \inf_{\lambda \geq \beta} \min\{u_\lambda^- - v_\lambda^-, u_\lambda^+ - v_\lambda^+\}, \sup_{\lambda \geq \beta} \max\{u_\lambda^- - v_\lambda^-, u_\lambda^+ - v_\lambda^+\} \right]$$

and it follows that

$$cl \bigcup_{\beta \geq \alpha} ([u]_\beta \ominus_{gH} [v]_\beta)$$

$$\subseteq \left[ \inf_{\beta \geq \alpha} \min\{u_\beta^- - v_\beta^-, u_\beta^+ - v_\beta^+\}, \sup_{\beta \geq \alpha} \max\{u_\beta^- - v_\beta^-, u_\beta^+ - v_\beta^+\} \right]$$

Let us consider now

$$cl \bigcup_{\beta \geq \alpha} ([u]_\beta \ominus_{gH} [v]_\beta)$$

$$= cl \bigcup_{\beta \geq \alpha} \left[ \min\{u_\beta^- - v_\beta^-, u_\beta^+ - v_\beta^+\}, \max\{u_\beta^- - v_\beta^-, u_\beta^+ - v_\beta^+\} \right].$$

For any $n \geq 1$, there exist

$$a_n \in \{u_\beta^- - v_\beta^-, u_\beta^+ - v_\beta^+ : \beta \geq \alpha\}$$

such that

$$\inf_{\beta \geq \alpha} \min\{u_\beta^- - v_\beta^-, u_\beta^+ - v_\beta^+\} > a_n - \frac{1}{n}.$$

Also there exist

$$b_n \in \{u_\beta^- - v_\beta^-, u_\beta^+ - v_\beta^+ : \beta \geq \alpha\}$$

such that

$$\sup_{\beta \geq \alpha} \max\{u_\beta^- - v_\beta^-, u_\beta^+ - v_\beta^+\} < b_n + \frac{1}{n}.$$

We have

$$cl \bigcup_{\beta \geq \alpha} ([u]_\beta \ominus_{gH} [v]_\beta) \supseteq [a_n, b_n], \forall n \geq 1$$

and we obtain

$$cl \bigcup_{\beta \geq \alpha} ([u]_\beta \ominus_{gH} [v]_\beta) \supseteq \bigcup_{n \geq 1} [a_n, b_n] \supseteq \left( \lim_{n \to \infty} a_n, \lim_{n \to \infty} b_n \right)$$

and finally

$$cl \bigcup_{\beta \geq \alpha} ([u]_\beta \ominus_{gH} [v]_\beta)$$

$$\supseteq \left[ \inf_{\beta \geq \alpha} \min\{u_\beta^- - v_\beta^-, u_\beta^+ - v_\beta^+\}, \sup_{\beta \geq \alpha} \max\{u_\beta^- - v_\beta^-, u_\beta^+ - v_\beta^+\} \right].$$

The conclusion

$$\left[\inf_{\beta\geq\alpha}\min\{u_\beta^- - v_\beta^-, u_\beta^+ - v_\beta^+\}, \sup_{\beta\geq\alpha}\max\{u_\beta^- - v_\beta^-, u_\beta^+ - v_\beta^+\}\right]$$

$$= cl\bigcup_{\beta\geq\alpha}([u]_\beta \ominus_{gH} [v]_\beta)$$

of the proposition follows.                                          ∎

**Proposition 5.20.** *For any fuzzy numbers $u, v \in \mathbb{R}_{\mathcal{F}}$ the g-difference $u \ominus_g v$ exists and it is a fuzzy number.*

**Proof.** According to the previous result, if we denote $w^- = (u \ominus_g v)^-$ and $w^+ = (u \ominus_g v)^+$ we have

$$w^-(\alpha) = \inf_{\beta\geq\alpha}\min\{u_\beta^- - v_\beta^-, u_\beta^+ - v_\beta^+\}$$

$$\leq w^+(\alpha) = \sup_{\beta\geq\alpha}\max\{u_\beta^- - v_\beta^-, u_\beta^+ - v_\beta^+\}.$$

Obviously $w^-$ is bounded and non-decreasing while $w^+$ is bounded non-increasing. Also, $w^-, w^+$ are left continuous on $(0, 1]$, since $u^- - v^-, u^+ - v^+$ are left continuous on $(0, 1]$ and they are right continuous at 0 since so are the functions $u^- - v^-, u^+ - v^+$.                         ∎

**Example 5.21.** *Consider the trapezoidal $u = (2, 3, 5, 6)$ and the triangular $v = (0, 4, 8)$ fuzzy numbers. It is easy to see that their gH-difference does not exists. Indeed, if we suppose the contrary then we need to have*

$$u_\alpha^- - v_\alpha^- \leq u_\alpha^+ - v_\alpha^+$$

*simultaneously for every $\alpha \in [0, 1]$ or*

$$u_\alpha^- - v_\alpha^- \geq u_\alpha^+ - v_\alpha^+$$

*simultaneously for every $\alpha \in [0, 1]$. But we observe that*

$$2 = u_0^- - v_0^- \geq u_0^+ - v_0^+ = -2,$$

*while*

$$-1 = u_1^- - v_0^- \leq u_0^+ - v_0^+ = 1.$$

*Since the same inequality should hold for every $\alpha \in [0, 1]$, we obtain a contradiction.*

*From the previous proposition the g-difference always exists so it exists and it is given as in Fig. 5.2.*

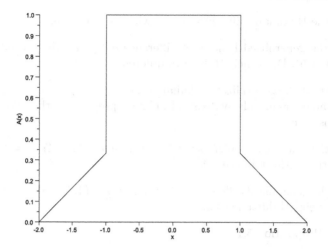

**Fig. 5.2** The g-difference of two fuzzy numbers

## 5.5  Problems

1. Let $u = (2, 5, 6, 7)$ and $v = (1, 4, 5)$ be a trapezoidal and a triangular fuzzy number. Calculate $u + v$, $u - v$, $-2u + v$, $u - u$ and $2v - v$.

2. Prove the properties in Theorem 5.8. (Hint: you can prove that the properties are true for every level set).

3. Find the product of the trapezoidal and triangular fuzzy numbers $u = (2, 5, 6, 8)$, $v = (1, 4, 5)$.

4. Find an expression for the level-sets of the product of two triangular fuzzy numbers $u = (a_1, b_1, c_1)$ and $v = (a_2, b_2, c_2)$, knowing that $a_1, a_2 > 0$.

5. Find an expression for the level-sets of the product of two Gaussian fuzzy numbers $u, v \in \mathbb{R}_{\mathcal{F}}$ with left endpoints of their level sets $u_0^-$, $v_0^- > 0$.

6. Find an expression for the level-sets of the product of two exponential fuzzy numbers $u, v \in \mathbb{R}_{\mathcal{F}}$ with left endpoints of their level sets $u_0^-$, $v_0^- > 0$.

7. Prove Proposition 5.12.

8. Prove that the cross product of two positive trapezoidal fuzzy numbers $u = (a_1, b_1, c_1, d_1)$ and $v = (a_2, b_2, c_2, d_2)$ is

$$u \odot v = (a_1 b_2 + a_2 b_1 - b_1 b_2, b_1 b_2, c_1 c_2, c_1 d_2 + c_2 d_1 - c_1 c_2).$$

Calculate the cross product with $u = (2, 3, 4, 5)$ and $v = (3, 4, 6, 8)$.

9. Find the Hukuhara difference $u \ominus v$, where $u = (2, 5, 6, 8)$, $v = (1, 4, 5)$.

10. Find the generalized Hukuhara difference $u \ominus_{gH} v$, where $u = (1, 2, 3)$, $v = (1, 3, 5)$. Does their Hukuhara difference exist?

11. Prove that the generalized Hukuhara difference $u \ominus_{gH} v$, of two triangular fuzzy numbers $u, v$ always exists. Find an expression for the gH-difference for this case.

12. Find the generalized difference $u \ominus_g v$, where $u = (0, 2, 4)$, $v = (0, 1, 2, 3)$. Does their gH-difference exist?

13. Find the generalized difference $u \ominus_g v$, where $u = (4, 5, 6, 8)$, $v = (0, 5, 10)$. Does their gH-difference exist?

14. Prove Proposition 5.17.

# 6
# Fuzzy Inference

Reasoning with imprecise information is one of the central topics of fuzzy logic. A fuzzy inference system consists of linguistic variables, fuzzy rules and a fuzzy inference mechanism. Linguistic variables allow us to interpret linguistic expressions in terms of fuzzy mathematical quantities. Fuzzy Rules are a set of rules that make association between typical input and output data sometimes in an intuitive way, or, on other occasions, in a data driven way. A fuzzy inference mechanism is able to model the process of approximate reasoning, through interpolation between the fuzzy rules. Of course good interpolations are also approximations, and in this way approximate reasoning is performed.

## 6.1   Linguistic Variables

**Definition 6.1.** *(Zadeh [156]) A **linguistic variable** is a quintuple*

$$(X, T, U, G, M)$$

*where*
  *$X$ is the name of the variable*
  *$T$ is the set of linguistic terms which can be values of the variable*
  *$U$ is the universe of discourse*
  *$G$ is a collection of syntax rules, grammar, that produces correct expressions in $T$.*
  *$M$ is a set of semantic rules that map $T$ into fuzzy sets in $U$.*

B. Bede: *Mathematics of Fuzzy Sets and Fuzzy Logic*, STUDFUZZ 295, pp. 79–103.
DOI: 10.1007/978-3-642-35221-8_6      © Springer-Verlag Berlin Heidelberg 2013

We can see from the previous definition that a linguistic variable works as a dictionary that translates linguistic terms into fuzzy sets. Often we use the term linguistic variable for a given value of the linguistic variable and also, if confusion is avoided, we use the same term for the fuzzy set that is associated to it, i.e., if $A$ is a fuzzy set that is associated through a semantic rule to an instance of a linguistic variable, then we say that $A$ is a linguistic variable.

**Example 6.2.** *We consider an example of a linguistic variable $(Age, T, U, G, M)$ where*

$X = Age$.

$T = \{young, \; very \; young, \; very \; very \; young, ...\}$

$U = [0, 100]$ *is the universe of discourse for age.*

$G :$ *The syntax rules can be expressed as follows: young $\in G$. If $x \in G$ then very $x \in G$.*

$M : T \to \mathcal{F}(X)$, $M(young) = u$, *where* $u = (0, 0, 18, 40)$.

$M(very^n \, young) = u^n(x)$.

As the previous example pointed out linguistic terms are sometimes composed of two parts. A fuzzy predicate (young, smart, small, tall, low) and modifiers (hedges) (very, likely, unlikely, extremely). Hedges may be interpreted as a composition between a given function and a basic membership function.

**Example 6.3.** *We consider an example linked to room temperature control $(Temperature, T, U, G, M)$ where*

$X = Temperature$.

$T = \{cold, \; very \; cold, ...cool, \; very \; cool, ...,...., hot, \; very \; hot,...\}$.

$U = [40, 100]$ *is the universe of discourse for temperature.*

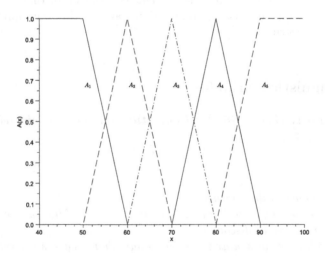

**Fig. 6.1** Example of fuzzy values of a linguistic variable

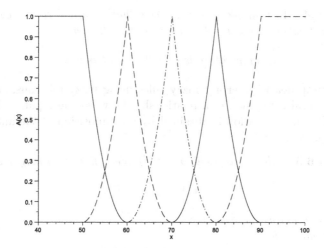

**Fig. 6.2** Modifiers with linguistic variables

*G : The syntax rules can be expressed as follows: cold, cool, just right, warm, hot $\in G$. If $x \in G$ then very $x \in G$. Let us observe that very warm$\neq$hot, they have different membership functions. Also, the expression very just right is correctly obtained by applying the hedge very to just right. It is a correct expression in our syntax however the linguistic term for it is not used in natural language.*

$M : T \rightarrow \mathcal{F}(X)$, $M(cold) = u_1$, where $u_1 = (40, 40, 50, 60)$, $M(cool) = u_2$, where $u_2 = (50, 60, 70)$, ..., $M(hot) = u_5$, where $u_5 = (80, 90, 100, 100)$.

$M(very^n cold) = u_1^n(x)$,..., $M(very^n hot) = u_5^n(x)$. See Figures 6.1, 6.2.

## 6.2 Fuzzy Rules

**Fuzzy rules** (fuzzy if-then rules) are able to model expert opinion or commonsense knowledge often expressed in linguistic terms. The intuitive association that exists between given typical input data and typical output data is hard to be described in a mathematically correct way, because of the uncertain, often subjective nature of this information. Fuzzy rules are tools that are able to model and use such knowledge.

A fuzzy rule is a triplet $(A, B, R)$ that consists of an antecedent $A \in \mathcal{F}(X)$, a consequence $B \in \mathcal{F}(X)$ that are linguistic variables, linked through a fuzzy relation $R \in \mathcal{F}(X \times Y)$.

Using fuzzy sets a fuzzy rule is written as follows:

$$\text{If } x \text{ is } A \text{ then } y \text{ is } B$$

**Example 6.4.** *An example of a fuzzy rule that naturally can be considered in the room temperature control problem is the following:*

*If temperature is cold then heat is high*

It is a natural idea to interpret fuzzy rules using fuzzy relations, since any rule is a natural expression of a relationship between the input and output variable. The fuzzy relation is obtained by a composition of the antecedent and consequence.

**Definition 6.5.** *(Mamdani-Assilian [107]) We define the **fuzzy rule***

*If x is A then y is B.*

*as a fuzzy relation as follows*
  *(i) Mamdani rule:*
$$R_M(x, y) = A(x) \wedge B(y);$$

*(ii) Larsen rule:*
$$R_L(x, y) = A(x) \cdot B(y);$$

*(iii) t-norm rule:*
$$R_T(x, y) = A(x)TB(y),$$

*with T being an arbitrary t-norm.*
  *(iv) Gödel rule:*
$$R_G(x, y) = A(x) \to B(y),$$

*with → being Gödel implication;*
  *(v) Gödel residual rule:*

$$R_R(x, y) = A(x) \to_T B(y)$$

*with →$_T$ being a residual implication with a given t-norm.*

**Example 6.6.** *An example of a fuzzy rule interpreted as a fuzzy relation is represented in Fig. 6.3. Two Gaussian fuzzy sets $A : [0, 1] \to [0, 1]$ and $B : [0, 1] \to [0, 1]$ are considered with $\bar{x}_1 = 0.6$ and $\sigma_1 = 0.1$ being the mean and spread of A, while $\bar{x}_2 = 0.4$ and $\sigma_2 = 0.1$ are the mean and the spread of B.*

**Remark 6.7.** *In many applications a fuzzy rule will have several antecedents that are used in conjunction to build our fuzzy rule. For example a more complex fuzzy rule can be considered*

*If x is A and y is B then z is C.*

*In this case the antecedents are naturally combined into a fuzzy relation*

$$D(x, y) = A(x) \wedge B(y)$$

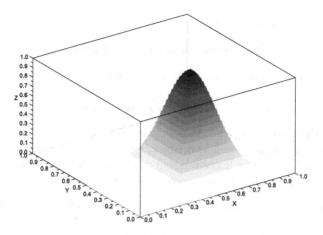

**Fig. 6.3** A fuzzy rule interpreted as a fuzzy relation

*that is regarded as a fuzzy set on its own. Then the fuzzy rule uses the antecedent D. For example, in this case the Mamdani rule will be*

$$R_M(x, y, z) = A(x) \wedge B(y) \wedge C(z).$$

*Between the antecedents we can use a general fuzzy conjunction (t-norm) but we do not have usually an implication between the antecedents since they are not in a cause effect relation with each-other. So, for example when a fuzzy rule with more antecedents is transformed into a fuzzy relation we can define the Gödel rule as*

$$R_G(x, y, z) = A(x) \wedge B(y) \to C(z).$$

## 6.3   Fuzzy Rule Base

A single fuzzy relation is barely enough to make an informed decision. Often in applications we will have a fuzzy knowledge base or **fuzzy rule base**, i.e. a finite collection of fuzzy rules.

**Example 6.8.** *Fuzzy rule bases are able to express expert or commonsense knowledge*

> *If temperature is cold then heat is high*

> *If temperature is cool then heat is low*

> *If temperature is just right then heat is off*

*If temperature is warm then cool is low*

*If temperature is hot then cool is high.*

*A fuzzy rule base can be expressed as a sequence of fuzzy rules*

*If temp is $A_i$ then heat/cool is $B_i$, $i = 1, ..., n$.*

We saw in the previous section that it is natural to translate one fuzzy rule into a fuzzy relation. In a similar manner, a fuzzy rule base can be translated into a fuzzy relation. The fuzzy relation is obtained by a max-min type or a min→ type composition of the antecedents and consequences.

**Definition 6.9.** *(Mamdani-Assilian [107]) We define the* **fuzzy rule base**

*If $x$ is $A_i$ then $y$ is $B_i$, $i = 1, ..., n$.*

*as a fuzzy relation as follows:*
*(i) Mamdani rule base:*

$$R_M(x, y) = \bigvee_{i=1}^{n} A_i(x) \wedge B_i(y)$$

*(ii) Larsen rule base:*

$$R_L(x, y) = \bigvee_{i=1}^{n} A_i(x) \cdot B_i(y)$$

*(iii) max-t-norm rule base:*

$$R_T(x, y) = \bigvee_{i=1}^{n} A_i(x) T B_i(y),$$

*with $T$ being an arbitrary t-norm*
*(iv) Gödel rule base:*

$$R_G(x, y) = \bigwedge_{i=1}^{n} A_i(x) \rightarrow B_i(y),$$

*with $\rightarrow$ being Gödel implication*
*(v) Gödel residual rule base:*

$$R_R(x, y) = \bigwedge_{i=1}^{n} A_i(x) \rightarrow_T B_i(y),$$

*with $\rightarrow_T$ being a residual implication with a given t-norm $T$.*

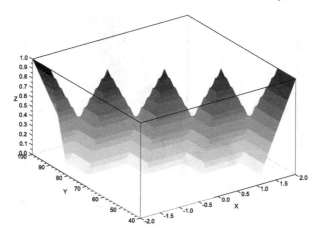

**Fig. 6.4** The Mamdani rule base

**Remark 6.10.** *As individual rules may have more antecedents linked through conjunctions we can have the same situation for a fuzzy rule base*

*If x is $A_i$ and y is $B_i$ then z is $C_i, i = 1, ..., n$.*

*In this case the antecedents are combined into a fuzzy relation $A_i(x) \wedge B_i(y)$. Then the Mamdani rule will be*

$$R_M(x, y, z) = \bigvee_{i=1}^{n} A_i(x) \wedge B_i(y) \wedge C_i(z).$$

*Between the antecedents we do not have an implication, instead we will have conjunction or eventually a t-norm. So, when a fuzzy rule base with more antecedents is translated into a fuzzy relation we obtain the Gödel rule base as*

$$R_G(x, y, z) = \bigwedge_{i=1}^{n} A_i(x) \wedge B_i(y) \to C_i(z).$$

**Example 6.11.** *Considering the temperature control example described above, the output can be scaled on the $[-2, 2]$ interval ($-2$ being the strongest cooling and 2 stands for the strongest heating. The consequences $B_1, ..., B_5$ are triangular fuzzy numbers $B_1 = (-2, -2, -1)$, $B_2 = (-2, -1, 0)$, $B_3 = (-1, 0, 1)$, $B_4 = (0, 1, 2)$, $B_5 = (1, 2, 2)$. The Mamdani rule base can be represented as in Fig. 6.4. The Gödel rule base of the same example is shown in Fig. 6.5.*

**Remark 6.12.** *We can see that there is not a unique way to interpret a fuzzy rule or a fuzzy rule base. This seems to be a drawback of fuzzy inference systems at first sight. We can turn this into an advantage because it gives us*

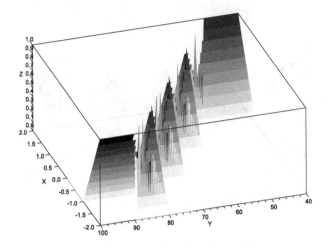

**Fig. 6.5** The Gödel rule base

*increased flexibility. Indeed, mainly in practical problems, it is an advantage to have more flexibility and to be able to select the method that best fits our problem.*

## 6.4   Fuzzy Inference

Fuzzy inference is the process of obtaining a conclusion for a given input that was possibly never encountered before. The basic rule (law) for a fuzzy inference system is the **compositional rule of inference** (Zadeh [155]) It is based on the classical rule of Modus Ponens. Let us recall first the classical Modus Ponens of Boolean logic:

$$premise: \text{ if } p \text{ then } q$$

$$fact: p$$

$$conclusion: q$$

Given a fuzzy rule or a fuzzy rule base $R \in \mathcal{F}(X \times Y)$, the compositional rule of inference is a function $F : \mathcal{F}(X) \to \mathcal{F}(Y)$ determined through a composition $B' = F(A') = A' * R$, with $* : \mathcal{F}(x) \times \mathcal{F}(X \times Y) \to \mathcal{F}(Y)$ being a composition of fuzzy relations.

The compositional rule of inference consists of a

$$premise: \text{ if } x \text{ is } A_i \text{ then } y \text{ is } B_i, i = 1, ..., n$$

$$fact: x \text{ is } A'$$

$$conclusion: y \text{ is } B'$$

**Definition 6.13.** *(Mamdani-Assilian [107], see also e.g. Fullér [64], Fullér [63]) We define a **fuzzy inference** based on a composition law as follows:*
*(i) Mamdani Inference:*

$$B'(y) = A' \circ R(x,y) = \bigvee_{x \in X} A'(x) \wedge R(x,y)$$

*(ii) Larsen inference:*

$$B'(y) = A' \circ_L R(x,y) = \bigvee_{x \in X} A'(x) \cdot R(x,y)$$

*(iii) Generalized modus ponens or t-norm-based inference:*

$$B'(y) = A' \circ_T R(x,y) = \bigvee_{x \in X} A'(x) T R(x,y)$$

*with T being an arbitrary t-norm.*
*(iv) Gödel Inference*

$$B'(y) = A' \triangleleft R(x,y) = \bigwedge_{x \in X} A'(x) \rightarrow R(x,y)$$

*with → being Gödel implication.*
*(v) Gödel residual inference*

$$B'(y) = A' \triangleleft_T R(x,y) = \bigwedge_{x \in X} A'(x) \rightarrow_T R(x,y)$$

*with $\rightarrow_T$ being a residual implication with a given t-norm.*
*Throughout the definition, $R(x,y)$ is a fuzzy relation that is used for interpreting the fuzzy rule base in the premise.*

**Remark 6.14.** *We can combine the above rule bases with any inference technique given in the previous definition. The combination that is most closely matching our real-world problem can be chosen or can be adaptively calculated.*

**Remark 6.15.** *As previously discussed we may have a fuzzy rule base with more antecedents, so we also need a fuzzy inference system to be able to deal with more antecedents. This is possible to be done by using the same strategy to combine the premises as the one used for the antecedents in the fuzzy rule base.*
*For example if we have a fuzzy inference system of the following form*

$$premise: \text{ If } x \text{ is } A_i \text{ and } y \text{ is } B_i \text{ then } z \text{ is } C_i$$

$$fact: \ x \text{ is } A' \text{ and } y \text{ is } B'$$

$$conclusion \ z \text{ is } C'$$

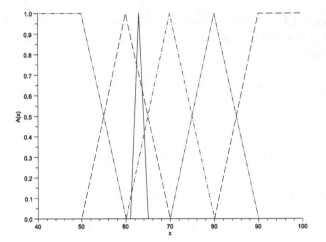

**Fig. 6.6** The input of a fuzzy inference system

*where $C'$ is determined based on a composition as for example*

$$C'(z) = (A' \wedge B') \circ R(x, y, z) = \bigvee_{x \in X, y \in Y} A'(x) \wedge B'(y) \wedge R(x, y, z)$$

*(for a Mamdani inference) and*

$$C'(z) = (A' \wedge B') \vartriangleleft R(x, y, z)$$
$$= \bigwedge_{x \in X, y \in Y} A'(x) \wedge B'(y) \to R(x, y, z)$$

*(for Gödel inference).*

*The fuzzy relation $R(x, y, z)$ is being used here to interpret the fuzzy rule base in the premise.*

**Example 6.16.** *To build a fuzzy inference system that solves the room temperature control problem discussed in the previous sections we can use e.g. a Mamdani or Gödel inference together with a Mamdani or a Gödel rule base. In Fig. 6.6 the input of the system is represented.*

## 6.5  The Interpolation Property of a Fuzzy Inference System

A natural property to be required for a fuzzy inference system is the following **interpolation property** (see Fullér [64]).

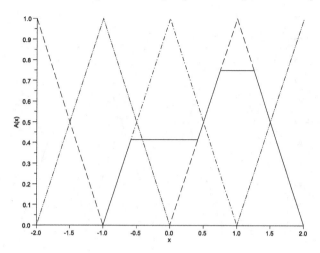

**Fig. 6.7** The output of a Mamdani inference system with Mamdani rule base

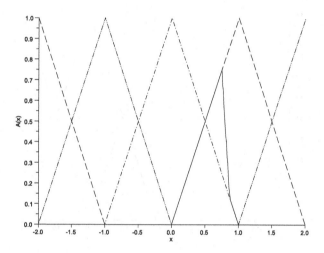

**Fig. 6.8** Output of a Mamdani fuzzy inference system with Gödel rule base

If the input of the system coincides with the antecedent of a fuzzy rule, then the output should coincide with the consequence that corresponds to the given antecedent through the fuzzy rule.

Let us consider first systems based on a single fuzzy rule. We saw that we can consider any combination of a fuzzy rule with a fuzzy inference, however the combinatorial nature of the problem does not permit an exhaustive analysis. So, we will consider some of the typical pairs of fuzzy rules and fuzzy inferences.

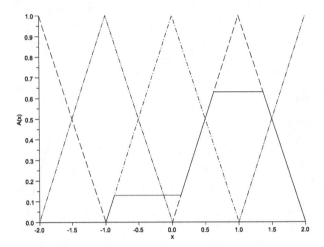

**Fig. 6.9** Output of a Gödel fuzzy inference system with Mamdani rule base

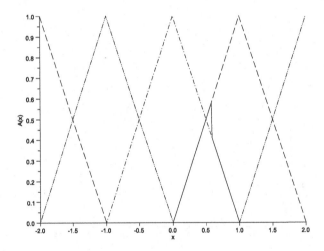

**Fig. 6.10** Output of a Gödel fuzzy inference system with Gödel rule base

Let us start by considering the Mamdani inference with Mamdani rule. The system is described by

$$B'(y) = \bigvee_{x \in X} A'(x) \wedge R(x, y), \ R(x, y) = A(x) \wedge B(y)$$

The interpolation property says that when the input of the system coincides with the antecedent $(A' = A)$ then the output has to be coincident with the consequence $(B' = B)$, i.e., our fuzzy inference is a generalization of the Modus Ponens of classical logic.

**Proposition 6.17.** *The Mamdani inference with Mamdani rule*

$$B'(y) = \bigvee_{x \in X} A'(x) \wedge R(x,y), \;\; R(x,y) = A(x) \wedge B(y),$$

*such that there exists an $x_0 \in X$ with $A(x_0) = 1$, satisfies the interpolation property, i.e., if $A' = A$, then $B' = B$.*

**Proof.** We have

$$B'(y) = \bigvee_{x \in X} A(x) \wedge R(x,y) = \bigvee_{x \in X} A(x) \wedge A(x) \wedge B(y)$$

$$= \bigvee_{x \in X} (A(x) \wedge B(y)) = \left( \bigvee_{x \in X} A(x) \right) \wedge B(y).$$

Since $A$ is normalized we have $\bigvee_{x \in X} A(x) = 1$ and then $B'(y) = 1 \wedge B(y) = B(y)$. ∎

**Remark 6.18.** *The hypothesis of the previous proposition can be weakened from $A$ being normal to $\bigvee_{x \in X} A(x) \geq \bigvee_{y \in X} B(y)$.*

Let us consider the Mamdani inference with Larsen rule.

**Proposition 6.19.** *The Mamdani inference with Larsen rule*

$$B'(y) = \bigvee_{x \in X} A'(x) \wedge R(x,y), \;\; R(x,y) = A(x) \cdot B(y)$$

*such that there exists an $x_0 \in X$ with $A(x_0) = 1$ satisfies the interpolation property, i.e., if $A' = A$ then $B' = B$.*

**Proof.** We leave the proof as an exercise to the reader. ∎

We can generalize the previous two propositions for an arbitrary t-norm.

**Proposition 6.20.** *The Mamdani inference with t-norm rule*

$$B'(y) = \bigvee_{x \in X} A'(x) \wedge R(x,y), \;\; R(x,y) = A(x)TB(y)$$

*such that there exists an $x_0 \in X$ with $A(x_0) = 1$ satisfies the interpolation property, i.e., if $A' = A$ then $B' = B$.*

**Proof.** From the properties of a t-norm we immediately get $A(x)TB(y) \leq A(x)$. We have

$$B'(y) = \bigvee_{x \in X} A(x) \wedge R(x,y)$$

$$= \bigvee_{x \in X} A(x) \wedge (A(x)TB(y)) = \bigvee_{x \in X} A(x)TB(y)$$

$$= \left( \bigvee_{x \in X} A(x) \right) T B(y) = 1 T B(y) = B(y).$$

∎

Let us consider the Mamdani inference with Gödel rule

$$B'(y) = \bigvee_{x \in X} A'(x) \wedge R(x,y), \ \ R(x,y) = A(x) \rightarrow B(y).$$

**Remark 6.21.** *We observe here that we can interpret the interpolation property of a Mamdani inference with Gödel rule as a fuzzy relational equation. Indeed, the interpolation property can be written as $A \circ R = B$. This equation, by Sanchez Theorem 3.24 has a solution if and only if $R(x,y) = A^{-1} \triangleleft B$ is a solution, and in this case it is the greatest solution $(A(x) = A(1,x), B(y) = B(1,y)$ are interpreted here as row vectors then*

$$R(x,y) = A^{-1} \triangleleft B(x,y)$$

$$= \bigwedge_{i=1} A(x,1) \rightarrow B(1,y) = A(x) \rightarrow B(y)$$

*becomes the Gödel implication. As a conclusion, the fuzzy relational equation is solvable if and only if the inference system with Mamdani inference and Gödel rule has the interpolation property.*

**Proposition 6.22.** *The Mamdani inference with Gödel rule*

$$B'(y) = \bigvee_{x \in X} A'(x) \wedge R(x,y), \ \ R(x,y) = A(x) \rightarrow B(y)$$

*such that $A$ is continuous and there exists an $x_0 \in X$ such that $A(x_0) = 1$ fulfills the interpolation property, i.e., if $A' = A$ then $B' = B$.*

**Proof.** We have

$$B'(y) = \bigvee_{x \in X} A(x) \wedge R(x,y)$$

$$= \bigvee_{x \in X} A(x) \wedge (A(x) \rightarrow B(y)).$$

Let

$$U_y = \{x \in X | A(x) \le B(y)\}.$$

Then we have

$$= \bigvee_{x \in U_y} A(x) \wedge (A(x) \rightarrow B(y)) \vee \bigvee_{x \notin U_y} A(x) \wedge (A(x) \rightarrow B(y))$$

$$= \bigvee_{x \in U_y} A(x) \wedge 1 \vee \bigvee_{x \notin U_y} A(x) \wedge B(y)$$

$$= \bigvee_{x \in U_y} A(x) \vee \bigvee_{x \notin U_y} B(y) = B(y) \vee B(y) = B(y).$$

We have used here the property that if $A$ is continuous then it attains its boundary which is $B(y)$ within $U_y$. ■

Let us consider the Gödel inference with Mamdani rule. The system is described by

$$B'(y) = \bigwedge_{x \in X} A'(x) \to R(x, y), \quad R(x, y) = A(x) \wedge B(y).$$

**Remark 6.23.** *We observe here that we can interpret the interpolation property of a Gödel inference with Mamdani rule as a fuzzy relational equation $A \triangleleft R = B$. This equation, by Miyakoshi-Shimbo Theorem 3.27 has a solution if and only if $R(x, y) = A^{-1} \circ B$ is a solution, and in this case it is the least solution and then*

$$R(x, y) = A^{-1} \circ B(x, y) = \bigvee_{i=1} A(x, 1) \wedge B(1, y) = A(x) \wedge B(y)$$

*becomes the Mamdani rule. As a conclusion, the equation is solvable if and only if the inference system with Mamdani inference and Gödel rule has the interpolation property.*

**Proposition 6.24.** *The Gödel inference with Mamdani rule*

$$B'(y) = \bigwedge_{x \in X} A'(x) \to R(x, y), \quad R(x, y) = A(x) \wedge B(y)$$

*with $A \in \mathcal{F}$ such that there exists an $x_0 \in X$ such that $A(x_0) = 1$ possesses the interpolation property, i.e., if $A' = A$ then $B' = B$.*

**Proof.** We have

$$B'(y) = \bigwedge_{x \in X} A(x) \to R(x, y) = \bigwedge_{x \in X} A(x) \to (A(x) \wedge B(y)).$$

Let $U_y = \{x \in X | A(x) \leq B(y)\}$. Then we have

$$= \bigwedge_{x \in U_y} A(x) \to (A(x) \wedge B(y)) \wedge \bigwedge_{x \notin U_y} A(x) \to (A(x) \wedge B(y))$$

$$= \bigwedge_{x \in U_y} A(x) \to A(x) \wedge \bigwedge_{x \notin U_y} A(x) \to B(y)$$

$$= 1 \wedge \bigwedge_{x \notin U_y} B(y) = B(y).$$

The infimum $\bigwedge_{x \notin U_y} B(y)$ exists since the normality of $A$ implies that $X - U_y$ is nonempty whenever $B(y) < 1$.    ∎

As we have discussed before we can rarely rely on one single fuzzy rule when designing a fuzzy inference system, instead, in most of the situations we have a rule base. So we will analyze the interpolation property of a fuzzy inference system together with a fuzzy rule base.

**Proposition 6.25.** *Let us consider the Mamdani inference with Mamdani rule base*

$$B'(y) = \bigvee_{x \in X} A'(x) \wedge R(x, y), \ R(x, y) = \bigvee_{i=1}^{n} A_i(x) \wedge B_i(y)$$

*such that*
*(i) for every $i = 1, ..., n$ there exists an $x_i \in X$ such that $A_i(x_i) = 1$;*
*(ii) any two antecedents $A_i, A_j$, $i \neq j$ have disjoint supports*
*Then the interpolation property is verified, i.e., if $A' = A_j$ then $B' = B_j$.*

**Proof.** We have

$$B'(y) = \bigvee_{x \in X} A_j(x) \wedge R(x, y) = \bigvee_{x \in X} A_j(x) \wedge \bigvee_{i=1}^{n} A_i(x) \wedge B_i(y)$$

$$= \bigvee_{i=1}^{n} \bigvee_{x \in X} A_j(x) \wedge A_i(x) \wedge B_i(y) = \bigvee_{i=1}^{n} B_i(y) \wedge \bigvee_{x \in X} A_j(x) \wedge A_i(x)$$

$$= B_j(y) \wedge \bigvee_{x \in X} A_j(x) \wedge A_j(x) = B_j(y).$$

The fact that $A_i$ and $A_j$ have disjoint support gives $\bigvee_{x \in X} A_j(x) \wedge A_i(x) = \delta_{ij}$ ($\delta_{ij}$ being Kronecker delta).    ∎

Next we consider the Mamdani inference with Larsen rule base.

**Proposition 6.26.** *Let us consider the Mamdani inference with Larsen rule base*

$$B'(y) = \bigvee_{x \in X} A'(x) \wedge R(x, y), \ R(x, y) = \bigvee_{i=1}^{n} A_i(x) \cdot B_i(y)$$

*such that*
*(i) for every $i = 1, ..., n$ there exists an $x_i \in X$ such that $A_i(x_i) = 1$;*
*(ii) any two antecedents $A_i, A_j$, $i \neq j$ have disjoint supports*
*Then the interpolation property is verified, i.e., if $A' = A_j$ then $B' = B_j$.*

**Proof.** The proof is left as an exercise.    ∎

**Remark 6.27.** *The requirement that the antecedents have disjoint support in the previous two propositions is too strong. This requirement gives that if $A'$ has its support disjoint from any of the antecedents, the output of the fuzzy system is 0. If we do not require a disjoint support then we can prove the following result.*

**Proposition 6.28.** *Let us consider the Mamdani inference with Mamdani rule base*

$$B'(y) = \bigvee_{x \in X} A'(x) \wedge R(x,y), \ R(x,y) = \bigvee_{i=1}^{n} A_i(x) \wedge B_i(y)$$

*such that for every $i = 1, ..., n$ there exists an $x_i \in X$ such that $A_i(x_i) = 1$. Under this condition, if $A' = A_j$ then $B' \geq B_j$.*

**Proof.** The proof is left as an exercise.    ∎

Let us consider now a Mamdani inference with Gödel rule base. The system is described by

$$B'(y) = \bigvee_{x \in X} A'(x) \wedge R(x,y), \ R(x,y) = \bigwedge_{i=1}^{n} A_i(x) \rightarrow B_i(y).$$

**Remark 6.29.** *We observe here that we can interpret the interpolation property of a Mamdani inference with Gödel rule base as a fuzzy relational equation. Indeed, the interpolation property can be written as $A_i \circ R = B_i$, $i = 1, ..., n$. If $A_i(x) = A(i,x)$, $B_i(y) = B(i,y)$, $i = 1, ..., n$ are interpreted as fuzzy relations we can rewrite our equation as $A \circ R = B$. This equation has a solution if and only if $R(x,y) = A^{-1} \triangleleft B$ is a solution, and in this case it is the greatest solution*

$$R(x,y) = A^{-1} \triangleleft B(x,y) = \bigwedge_{i=1}^{n} A(x,i) \rightarrow B(i,y)$$

$$= \bigwedge_{i=1}^{n} A_i(x) \rightarrow B_i(y).$$

*As a conclusion, the equation is solvable if and only if the inference system with Mamdani inference and Gödel rule base has the interpolation property.*

**Proposition 6.30.** *Let us consider a Mamdani inference with Gödel rule base*

$$B'(y) = \bigvee_{x \in X} A'(x) \wedge R(x,y), \ R(x,y) = \bigwedge_{i=1}^{n} A_i(x) \rightarrow B_i(y).$$

*(i) If $A_i$ are normal and continuous and*

*(ii) if the core of any $A_i$ and the support of any $A_j$ are disjoint when $i \neq j$,*

*then the Mamdani inference with Gödel rule base satisfies the interpolation property: if $A' = A_j$ then $B' = B_j$.*

**Proof.** We have

$$B'(y) = \bigvee_{x \in X} A_j(x) \wedge \bigwedge_{i=1}^{n} A_i(x) \rightarrow B_i(y)$$

$$\leq \bigvee_{x \in X} A_j(x) \wedge (A_j(x) \rightarrow B_j(y)) \leq B_j(y),$$

where the last equality is obtained from the property $x \wedge (x \rightarrow y) \leq y$. Also,

$$B'(y) = \bigvee_{x \in X} A_j(x) \wedge \bigwedge_{i=1}^{n} A_i(x) \rightarrow B_i(y)$$

$$= \bigvee_{x \in X} \bigwedge_{i=1}^{n} A_j(x) \wedge (A_i(x) \rightarrow B_i(y))$$

$$\geq \bigwedge_{i=1}^{n} A_j(\bar{x}) \wedge (A_i(\bar{x}) \rightarrow B_i(y)), \forall \bar{x} \in X.$$

We take $\bar{x} \in core(A_j)$ and $\bar{x} \notin supp(A_i)$ then

$$B'(y) \geq A_j(\bar{x}) \wedge (A_j(\bar{x}) \rightarrow B_j(y))$$

$$\wedge \bigwedge_{i \neq j} A_j(\bar{x}) \wedge (A_i(\bar{x}) \rightarrow B_i(y))$$

$$= 1 \wedge (1 \rightarrow B_j(y)) \wedge \bigwedge_{i \neq j} 1 \wedge (0 \rightarrow B_i(y))$$

$$= B_j(y) \wedge 1 = B_j(y).$$

Combining the two inequalities leads to the required conclusion.    ∎

Condition (ii) and continuity are not too strong requirements, however if released we still have an inequality.

**Proposition 6.31.** *Let us consider a Mamdani inference with Gödel rule base*

$$B'(y) = \bigvee_{x \in X} A'(x) \wedge R(x, y), \quad R(x, y) = \bigwedge_{i=1}^{n} A_i(x) \rightarrow B_i(y).$$

*If $A_i$ are normal and if $A' = A_j$ then $B' \leq B_j$.*

**Proof.** We leave the proof to the reader as an exercise.                    ∎

A rule base that has the antecedents fulfilling the property described in this theorem is said to be separated.

Let us consider the Gödel inference with Mamdani rule base. The system is described by

$$B'(y) = \bigwedge_{x \in X} A'(x) \to R(x,y), \ \ R(x,y) = \bigvee_{i=1}^{n} A_i(x) \wedge B_i(y).$$

**Remark 6.32.** *We observe here that we can interpret the interpolation property of a Gödel inference with Mamdani rule base as a fuzzy relational equation* $A_i \triangleleft R = B_i, i = 1, ..., n$ *or considering* $A(i,x) = A_i(x)$ *and* $B(i,x) = B_i(x)$ *we can write* $A \triangleleft R = B$. *This equation, by Miyakoshi-Shimbo theorem has a solution if and only if* $R(x,y) = A^{-1} \circ B$ *is a solution, and in this case it is the least solution and we also have*

$$R(x,y) = A^{-1} \circ B(x,y) = \bigvee_{i=1} A(x,i) \wedge B(i,y)$$

$$= \bigvee_{i=1}^{n} A_i(x) \wedge B_i(y).$$

*becomes the Gödel implication. As a conclusion, the equation is solvable if and only if the inference system with Mamdani inference and Gödel rule base has the interpolation property.*

**Proposition 6.33.** *Let us consider the Gödel inference with Mamdani rule base*

$$B'(y) = \bigwedge_{x \in X} A'(x) \to R(x,y), \ \ R(x,y) = \bigvee_{i=1}^{n} A_i(x) \wedge B_i(y).$$

*(i) If* $A_i$ *are normal and continuous*

*(ii) and if the core of any* $A_i$ *and the support of any* $A_j$ *are disjoint when* $i \neq j$,

*then the interpolation property holds i.e., if* $A' = A_j$ *then* $B' = B_j$.

**Proof.** We have

$$B'(y) = \bigwedge_{x \in X} A_j(x) \to R(x,y)$$

$$= \bigwedge_{x \in X} A_j(x) \to \bigvee_{i=1}^{n} A_i(x) \wedge B_i(y)$$

$$\geq \bigwedge_{x \in X} A_j(x) \to A_j(x) \wedge B_j(y) \geq B_j(y).$$

We take $\bar{x} \in core(A_j)$ and $\bar{x} \notin supp(A_i)$ then

$$B'(y) = \bigwedge_{x \in X} A_j(x) \to \bigvee_{i=1}^{n} A_i(x) \wedge B_i(y)$$

$$\leq A_j(\bar{x}) \to \bigvee_{i=1}^{n} A_i(\bar{x}) \wedge B_i(y)$$

$$= (A_j(\bar{x}) \to A_j(\bar{x}) \wedge B_j(y)) \vee \bigvee_{i \neq j} A_j(\bar{x}) \to A_i(\bar{x}) \wedge B_i(y)$$

$$(1 \to 1 \wedge B_j(y)) \vee \bigvee_{i \neq j} 1 \to 0 \wedge B_i(y) = B_j(y) \vee 0 = B_j(y).$$

∎

We can obtain one of the inequalities under very light conditions.

**Proposition 6.34.** *Let us consider the Gödel inference with Mamdani rule base*

$$B'(y) = \bigwedge_{x \in X} A'(x) \to R(x,y), \quad R(x,y) = \bigvee_{i=1}^{n} A_i(x) \wedge B_i(y).$$

*If $A_i$ are normal and if $A' = A_j$ then $B' \geq B_j$.*

**Proof.** The proof is left to the reader.                    ∎

**Proposition 6.35.** *Let us consider the Gödel residual inference with Mamdani rule base*

$$B'(y) = \bigwedge_{x \in X} A'(x) \to_T R(x,y), \quad R(x,y) = \bigvee_{i=1}^{n} A_i(x) \wedge B_i(y).$$

*(i) If $A_i$ are normal and continuous*
*(ii) and if the core of any $A_i$ and the support of any $A_j$ are disjoint when $i \neq j$,*
*  then the interpolation property holds i.e., if $A' = A_j$ then $B' = B_j$.*

**Proof.** We leave the proof to the reader as an exercise.        ∎

## 6.6   Example of a Fuzzy Inference System

As an application of a fuzzy inference system we consider a problem of Computing with Words (see Zadeh [157]).

Let us consider a problem with Lt. Columbo. The problem was proposed in the paper Dvorak-Novak [54], but with a different solution, not the one presented here. The problem's statement is as follows (see Dvorak-Novak [54]):

**Example 6.36.** *"Mr. John Smith has been shot dead in his house. He was found by his friend, Mr. Carry. Lt. Columbo suspects Mr. Carry to be the murderer.*

*Mr. Carry's testimony is the following: I have started from my home at about 6:30, arrived to John's house at about 7, found John dead and went immediately to the phone box to call police. They told me to wait and came immediately.*

*Lt. Columbo has found the following evidence about dead Mr. Smith: He had high quality suit with broken wristwatch stopped at 5:45. No evidence of strong strike on his body. Lt. Columbo touched engine of Mr. Carry's car and found it to be more or less cold."*

*To be able to analyze the problem we need to understand and implement (following the ideas in Dvorak-Novak [54]) commonsense knowledge. We will use fuzzy if then rules for this task. Let us start with describing how engine temperature depends on drive duration.*

1. *If drive duration is big and time stopped is small then engine is hot.*

2. *If drive duration is small then engine is cold.*

3. *If time stopped is big then engine is cold.*

*Another set of fuzzy rules concerns the wristwatch quality.*

1. *If suit quality is high then wristwatch quality is high.*

2. *If suit quality is low then wristwatch quality is low.*

*The last set of fuzzy rules concerns how likely the wristwatch is broken.*

1. *If wristwatch quality is high and strike is unlikely then broken is unlikely.*

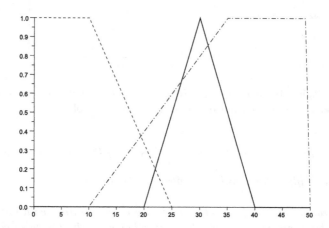

**Fig. 6.11** Drive duration small (dash), big (dash-dot) and about 30' from Mr. Carry's testimony (solid line)

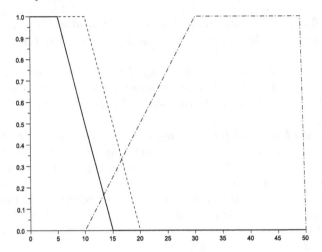

**Fig. 6.12** Time stooped small (dash), big (dash-dot) and the fuzzy set representing the fact that police came immediately (solid line)

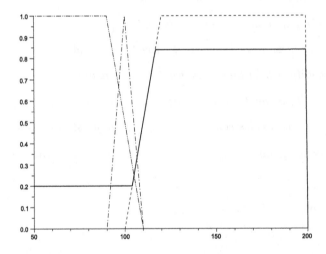

**Fig. 6.13** Engine Temperature cold (dash-double dot), hot (dash), more or less cold (dash-dot), and the output of the inference system (solid line)

*2. If wristwatch quality is low and strike is likely then broken is likely.*

*3. If strike is likely then broken is more or less likely.*

*The antecedents for the first set of rules are drive duration with antecedents small and big represented in Fig. 6.11, together with Mr. Carry's drive dura-tion of about 30' as he started from home at about 6:30 and arrived at John's house at about 7. The time since the car has stopped and cooled down is*

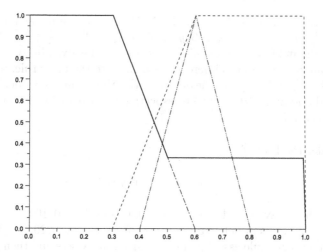

**Fig. 6.14** Watch broken unlikely (dash-dot), more or less likely (dash-double dot) likely (dash), and the output fuzzy set obtained by fuzzy inference (solid line)

*represented in Fig. 6.12. The engine temperature is represented in Fig 6.13. As an input considering Mr. Carry's drive duration and time stopped we can consider a fuzzy inference system to compute the fuzzy sets associated Mr. Carry's testimony and compare it with Lt. Columbo's observations. We used a Mamdani inference and a Mamdani rule base for simplicity but similar results are obtained using other inference systems. The output of the inference system (solid line in Fig. 6.13) represents the engine temperature inferred from Mr. Carry's testimony and it is compared with the engine temperature observed by Lt. Columbo. The result shows that according to Mr. Carry's testimony the engine temperature should be much higher than the one observed by Lt. Columbo, so Mr. Carry is lying. We have also considered a fuzzy inference system to determine how likely the wristwatch was broken at the time of John's death. We will skip some details and we show the output of the fuzzy system with Mamdani inference and rule base in Fig. 6.14. The evidence gathered by Lt. Columbo regarding qualitative estimates is used as input of a fuzzy inference system consisting of the last two set of fuzzy rules and we obtain that it is unlikely that the watch broke at the time of death of the victim. Since the watch is broken, this is considered indirect testimony by Mr. Carry and it turns out again that it contradicts the evidence, i.e., Mr. Carry is lying.*

## 6.7   Problems

1. Consider the fuzzy rule

$$\text{If } x \text{ is } A \text{ then } y \text{ is } B$$

where the fuzzy sets $A, B : \{1, 2, 3, 4, 5\} \rightarrow [0, 1]$ are defined as $A(x) = \frac{0.2}{1} + \frac{0.7}{2} + \frac{1}{3} + \frac{0.7}{4} + \frac{0.2}{5}$, $B(y) = \frac{0}{1} + \frac{0.5}{2} + \frac{0.8}{3} + \frac{1}{4} + \frac{0.5}{5}$. Find the Mamdani and Gödel fuzzy rules associated to the given fuzzy sets. Write the Mamdani and Gödel inference associated to the fuzzy rule. If $A'(x) = \frac{0.1}{1} + \frac{0.7}{2} + \frac{0.3}{3} + \frac{0.2}{4} + \frac{0}{5}$ find the output of the Mamdani and Gödel inference with Mamdani and Gödel rule respectively (each rule can be combined with each inference).

2. Consider the fuzzy rule

$$\text{If } x \text{ is } A \text{ then } y \text{ is } B$$

where the fuzzy sets $A : [0, 10] \rightarrow [0, 1]$ and $B : [0, 100] \rightarrow [0, 1]$ are triangular numbers defined as $A = (3, 5, 6)$ $B = (10, 20, 30)$. Find the Mamdani fuzzy rule associated to the given fuzzy sets and then graph it. Write the Mamdani inference associated to the fuzzy rule. If $A' = (2, 3, 4)$ then calculate the output of the Mamdani inference with Mamdani rule and then graph it.

3. Consider the fuzzy rule

$$\text{If } x \text{ is } A \text{ then } y \text{ is } B$$

where the fuzzy sets $A : [0, 10] \rightarrow [0, 1]$ and $B : [0, 100] \rightarrow [0, 1]$ are trapezoidal numbers defined as $A = (3, 5, 6, 8)$ $B = (10, 20, 30, 40)$. Find the Mamdani fuzzy rule associated to the given fuzzy sets and then graph it. Write the Mamdani inference associated to the fuzzy rule. If $A' = \chi_{\{4\}}$ is a singleton input then find the output of the Mamdani inference with Mamdani rule and then graph it.

4. Set up a fuzzy rule base and a fuzzy inference system for the problem of filling up a reservoir with water. We can consider for example three antecedents: water level is low, medium, high) and correspondingly three consequences debit is high, medium, low. Give the fuzzy if then rules, the antecedents and the consequences and explain what fuzzy rule base and inference you would use with them.

5. Set up a fuzzy inference system for an agent in a video-game that has to avoid the enemy. The antecedents that can be considered have to take into account that the enemy can be near or far form our agent and the enemy may be approaching, stationary or departing. The agent we are considering may be stationary, or moving away slow, moving away fast (from the enemy). Give the fuzzy if then rules, the antecedents and the consequences and explain what fuzzy rule base and inference you would use with them.

6. Set up a fuzzy inference system which has to decide on the tip at a restaurant depending on the quality of service and the quality of food. Write a set of fuzzy rules with antecedents and consequences, and a fuzzy inference. Consider a set of premises and find the corresponding conclusion.

7. Let us consider the Mamdani inference with Larsen rule. The system is described by

$$B'(y) = \bigvee_{x \in X} A'(x) \wedge R(x, y), \quad R(x, y) = A(x) \cdot B(y)$$

Prove that if $A$ is normal then the system has the interpolation property, i.e. if $A' = A$ then $B' = B$.

8. Let us consider the Mamdani inference with Larsen rule base. The system is described by

$$B'(y) = \bigvee_{x \in X} A'(x) \wedge R(x, y), \quad R(x, y) = \bigvee_{i=1}^{n} A_i(x) \cdot B_i(y)$$

Prove that if $A_i$ are normal and if they have disjoint supports, then the system has the interpolation property, i.e. if $A' = A_j$ then $B' = B_j$.

9. Let us consider the Mamdani inference with t-norm rule base

$$B'(y) = \bigvee_{x \in X} A'(x) \wedge R(x, y), \quad R(x, y) = \bigvee_{i=1}^{n} A_i(x) T B_i(y)$$

such that for every $i = 1, ..., n$ there exists an $x_i \in X$ such that $A_i(x_i) = 1$. Prove that the following property is verified: If $A' = A_j$ then $B' \geq B_j$

10. Let us consider a Mamdani inference with Gödel rule base

$$B'(y) = \bigvee_{x \in X} A'(x) \wedge R(x, y), \quad R(x, y) = \bigwedge_{i=1}^{n} A_i(x) \to B_i(y).$$

if $A_i$ are normal and if $A' = A_j$ then $B' \leq B_j$.

11. Let us consider the Gödel residual inference with Mamdani rule base

$$B'(y) = \bigwedge_{x \in X} A'(x) \to_T R(x, y), \quad R(x, y) = \bigvee_{i=1}^{n} A_i(x) \wedge B_i(y).$$

(i) If $A_i$ are normal and continuous
(ii) and if the core of any $A_i$ and the support of any $A_j$ are disjoint when $i \neq j$,
then the interpolation property holds i.e., if $A' = A_j$ then $B' = B_j$.

# 7

# Single Input Single Output Fuzzy Systems

In the previous chapter we have described in detail Fuzzy Inference Systems. These have fuzzy sets as inputs, and their output is a fuzzy set as well, so these systems work exclusively in a fuzzy setting. Often, in practical applications we need to be able to accept crisp inputs and also, the system needs to produce a crisp number for the output. Surely this is often a well defined classical functional relationship between inputs and outputs. Naturally raises the question why do we need fuzzy systems when we have a crisp relationship, crisp input and also a crisp output for a classical system. The reason for this fact lies in epistemic uncertainty. However we have a crisp relationship, this is often unknown or only partially known to us. As we will see in this chapter fuzzy systems can fill in the gaps and approximate any desired output with arbitrary precision. Approximation properties of fuzzy systems were widely investigated in the literature starting from B. Kosko's paper [97]. Since then many authors dealt with approximation properties of fuzzy systems Kosko [111], Li-Shi-Li [100], Li-Shi-Li [101] Tikk-Kóczy-Gedeon [146], Kóczy-Zorat [95]. Yet a fully constructive method providing also an error estimate is still missing. The approach in Li-Shi-Li [100], Li-Shi-Li [101] is constructive but not fully constructive, because it requires the knowledge of the inverse image of the function to be approximated. The present approach solves this problem. The difficulty in studying such systems is in the fact that they are nonlinear operators, based on max, min and $\to$ type operations. Tools to deal with similar problems were recently developed in Bede-Nobuhara-Dankova-Di Nola [22] and Bede-Coroianu-Gal [16]. The results in Section 7.3 are published for the first time in the present work.

B. Bede: *Mathematics of Fuzzy Sets and Fuzzy Logic*, STUDFUZZ 295, pp. 105–136.
DOI: 10.1007/978-3-642-35221-8_7      © Springer-Verlag Berlin Heidelberg 2013

Besides Mamdani approach (Mamdani Assilian [107]) there is another major direction, that of Takagi-Sugeno fuzzy systems (Sugeno [138]). Sugeno systems and their approximation capability were studied in Buckley [33], Ying [153], Celikyilmaz-Türksen [34], Sonbol-Fadali [132] etc. Again the constructive results are either missing from the literature or they are not accompanied by error estimates. In Section 7.5 we prove approximation properties for Takagi-Sugeno fuzzy systems and higher order Takagi-Sugeno fuzzy systems (see Celikyilmaz-Türksen [34]). The results of Section 7.4 are new, published in the present work for the first time. The error estimates presented here are simple and they open up new research directions and applications.

Combining these approximation properties of a fuzzy system with the ability to model linguistic expressions makes them very interesting to study and also very efficient in applications. This motivates us to emphasize the idea that the topic of Fuzzy Sets and Fuzzy Systems is an independent discipline within Mathematics, being a unique combination of ideas overlapping strongly with topics in Logic and Approximation Theory.

## 7.1    Structure of a SISO Fuzzy System

A **single input single output fuzzy system (SISO Fuzzy System)** uses a crisp input, fuzzifies it, maps it through a fuzzy inference system and the fuzzy output that is obtained is transformed into a crisp output. Often SISO Fuzzy Systems are used in a control problem and then they are called fuzzy controllers.

The diagram of a SISO fuzzy system is represented in Fig. 7.1.

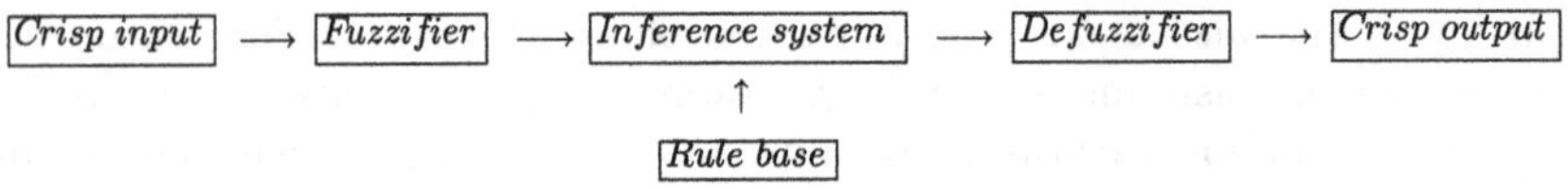

$$\boxed{\text{Crisp input}} \longrightarrow \boxed{\text{Fuzzifier}} \longrightarrow \boxed{\text{Inference system}} \longrightarrow \boxed{\text{Defuzzifier}} \longrightarrow \boxed{\text{Crisp output}}$$
$$\uparrow$$
$$\boxed{\text{Rule base}}$$

**Fig. 7.1** The components of a fuzzy controller

Let us discuss in what follows each of the components.

### 7.1.1    Fuzzification

Most of the systems use the most basic fuzzifier that is the canonical inclusion. If $x_0 \in X$ is a crisp input then the fuzzy set associated with it is the singleton fuzzy set $x_0$, given by the characteristic function

$$A'(x) = \chi_{\{x_0\}}(x) = \begin{cases} 1 & \text{if} \quad x = x_0 \\ 0 & \text{if} \quad x \neq x_0 \end{cases}.$$

Fuzzy Systems can take into account uncertainty at this step. If the input crisp value is influenced by uncertainties then these can be taken into account when building a non-singleton fuzzy input. In the present chapter we consider the inclusion fuzzifier.

## 7.1.2   Fuzzy Rule Base

The fuzzy rule base can be described as a fuzzy relation $R(x, y)$. For a fuzzy controller we can select among different types of fuzzy rule bases as described previously.

## 7.1.3   Fuzzy Inference

The fuzzy inference system that we consider can be of any type that have been discussed before (Mamdani, Gödel, Larsen etc.). Mamdani Inference gives

$$B'(y) = (R \circ A')(x) = \bigvee_{x \in X} A'(x) \wedge R(x, y),$$

where the fuzzy relation $R$ is the fuzzy rule base.

## 7.1.4   Defuzzification

Defuzzification is the final step in a fuzzy control algorithm. Based on the output of a fuzzy controller one has to give an estimate of the crisp quantity (a representative crisp element) for the output value of the SISO fuzzy system. In this case one has to use a defuzzification. There are many different defuzzification methods and based on the given application that we are working on, we can select a suitable defuzzification.

**Center of Gravity (COG).** The value selected is the center of gravity of the fuzzy set $u \in \mathcal{F}(X)$. More formally, we have

$$COG(u) = \frac{\int_W x \cdot u(x)dx}{\int_W u(x)dx},$$

where $W = supp(u)$. In the present work the Center of Gravity defuzzification is the most frequently used.

**Center of Area (COA).** If $u \in \mathcal{F}(\mathbb{R})$ then the number $COA(u)$ is defined as the point of the support of $u$ that divides the area under the membership function into two equal parts. If $COA(u) = a$ then $a$ satisfies

$$\int_{-\infty}^{a} u(x)dx = \int_{a}^{\infty} u(x)dx.$$

**Expected value and expected interval (EV and EVI).** If $u$ is a continuous fuzzy number then the expected value $EV(u)$ is given by

$$EV(u) = \frac{1}{2} \int_0^1 \left( u_r^- + u_r^+ \right) dr.$$

and the expected interval is

$$EVI(u) = \left[ \int_0^1 u_r^- dr, \int_0^1 u_r^+ dr \right].$$

The expected value is the midpoint of the expected interval.

**Mean of Maxima (MOM).** The mean of maxima defuzzifier selects the mean value of the points where the membership grade attains its maximum.

$$MOM(u) = \frac{\int_{x \in U} x dx}{\int_{x \in U} dx},$$

where $U = \{x \in X | u(x) = \max_{t \in X} u(t)\}$.

**Maximum criterion.** Sometimes, to avoid practical difficulties the choice of the defuzzification can be arbitrary fulfilling the max criterion i.e., we may select arbitrary $x \in U$ with

$$U = \{x \in X | u(x) = \max_{t \in X} u(t)\}.$$

## 7.2   Fuzzy Inference and Rule Base for a SISO Fuzzy System

Mamdani inference has simple expression on par with great computational and intuitive properties. Also, these were historically the systems used in the first fuzzy controllers. Using a Gödel inference has apparently some theoretical advantages over Mamdani inference because of their interpretation via fuzzy relational equations. The choice of Mamdani inference in SISO fuzzy systems is motivated by the following property.

**Proposition 7.1.** *If $A'(x) = \chi_{\{x_0\}}(x)$ is a crisp input of a fuzzy inference system with a given rule base $R(x, y)$, then the outputs of Mamdani, Larsen, t-norm, Gödel and Gödel residual inference systems coincide.*

**Proof.** Let

$$A'(x) = \begin{cases} 1 & \text{if} \quad x = x_0 \\ 0 & \text{if} \quad x \neq x_0 \end{cases}.$$

Then for the Mamdani inference we have

$$B'(y) = \bigvee_{x \in X} A'(x) \wedge R(x, y) = A'(x_0) \wedge R(x_0, y)$$

$$= 1 \wedge R(x_0, y) = R(x_0, y).$$

For Larsen inference we have

$$B'(y) = \bigvee_{x \in X} A'(x) \cdot R(x, y)$$

$$= A'(x_0) \cdot R(x_0, y) = R(x_0, y).$$

Then for the t-norm based inference we have

$$B'(y) = \bigvee_{x \in X} A'(x) T R(x, y) = A'(x_0) T R(x_0, y)$$

$$= 1 T R(x_0, y) = R(x_0, y).$$

For the Gödel inference we have

$$B'(y) = \bigwedge_{x \in X} A'(x) \to R(x, y)$$

$$= 1 \wedge A'(x_0) \to R(x_0, y)$$
$$= 1 \to R(x_0, y) = R(x_0, y).$$

For the Gödel residual inference we have

$$B'(y) = \bigwedge_{x \in X} A'(x) \to_T R(x, y)$$

$$= 1 \wedge A'(x_0) \to_T R(x_0, y)$$
$$= 1 \to_T R(x_0, y) = R(x_0, y).$$

∎

Similar result holds for the more general case of having more antecedents. In a SISO fuzzy system the inference system used does not have an impact on the output of the fuzzy system, so the output is determined solely by the rule base used. The rule base that we are using can still be chosen to be of Mamdani, Larsen, t-norm, Gödel or Gödel residual type. Since the inference system in the SISO case does not matter, the fuzzy rule base is the one that will be specified and it will decide the output of the system.

The following results show a simplification for the calculations of the output of a SISO fuzzy system and so, these systems will be computationally very inexpensive, making them widely applicable in real-time situations.

**Proposition 7.2.** *If $A'(x) = \chi_{\{x_0\}}(x)$ is a crisp input of a fuzzy inference system with a Mamdani rule base*

$$R(x, y) = \bigvee_{i=1}^{n} A_i(x) \wedge B_i(y)$$

*then the output of the system can be calculated as*

$$B'(y) = \bigvee_{i=1}^{n} \alpha_i \wedge B_i(y),$$

*where $\alpha_i = A_i(x_0)$ is the firing strength of the $i^{th}$ fuzzy rule.*

**Proof.** Let

$$A'(x) = \begin{cases} 1 & \text{if} \quad x = x_0 \\ 0 & \text{if} \quad x \neq x_0 \end{cases}.$$

Then we have

$$B'(y) = \bigvee_{x \in X} A'(x) \wedge \bigvee_{i=1}^{n} A_i(x) \wedge B_i(y)$$

$$= A'(x_0) \wedge \bigvee_{i=1}^{n} A_i(x_0) \wedge B_i(y)$$

$$= 1 \wedge \bigvee_{i=1}^{n} A_i(x_0) \wedge B_i(y) = \bigvee_{i=1}^{n} \alpha_i \wedge B_i(y),$$

where $\alpha_i = A_i(x_0)$. ∎

Similar results hold for the Larsen and t-norm based inferences.

**Proposition 7.3.** *If $A'(x)$ is a crisp input of a fuzzy inference system with a Larsen rule base*

$$R(x, y) = \bigvee_{i=1}^{n} A_i(x) \cdot B_i(y)$$

*then*

$$B'(y) = \bigvee_{i=1}^{n} \alpha_i \cdot B_i(y),$$

*where $\alpha_i = A_i(x_0)$ is the firing strength of the $i^{th}$ fuzzy rule.*

**Proof.** The proof is left as an exercise. ∎

**Proposition 7.4.** *If $A'(x)$ is a crisp input of a fuzzy inference system with a t-norm rule base*

$$R(x, y) = \bigvee_{i=1}^{n} A_i(x) T B_i(y)$$

*then*

$$B'(y) = \bigvee_{i=1}^{n} \alpha_i T B_i(y),$$

*where $\alpha_i = A_i(x_0)$ is the firing strength of the $i^{th}$ fuzzy rule.*

**Proof.** Let

$$A'(x) = \begin{cases} 1 & \text{if } x = x_0 \\ 0 & \text{if } x \neq x_0 \end{cases}.$$

Then we have

$$B'(y) = \bigvee_{x \in X} A'(x) \wedge \bigvee_{i=1}^{n} A_i(x) T B_i(y)$$

$$= \bigvee_{i=1}^{n} A_i(x_0) T B_i(y) = \bigvee_{i=1}^{n} \alpha_i T B_i(y)$$

with $\alpha_i = A_i(x_0)$. ∎

**Proposition 7.5.** *If $A'(x)$ is a crisp input of a fuzzy inference system with a Gödel rule base*

$$R(x, y) = \bigwedge_{i=1}^{n} A_i(x) \to B_i(y)$$

*then*

$$B'(y) = \bigwedge_{i=1}^{n} \alpha_i \to B_i(y),$$

*where $\alpha_i = A_i(x_0)$ is the firing strength of the $i^{th}$ fuzzy rule.*

**Proof.** Let

$$A'(x) = \begin{cases} 1 & \text{if } x = x_0 \\ 0 & \text{if } x \neq x_0 \end{cases}.$$

Then we have

$$B'(y) = \bigvee_{x \in X} A'(x) \wedge \bigwedge_{i=1}^{n} A_i(x) \to B_i(y)$$

$$= \bigwedge_{i=1}^{n} A'(x_0) \wedge (A_i(x_0) \to B_i(y))$$

$$= \bigwedge_{i=1}^{n} A_i(x_0) \to B_i(y) = \bigwedge_{i=1}^{n} \alpha_i \to B_i(y),$$

where $\alpha_i = A_i(x_0)$. ∎

**Proposition 7.6.** *If $A'(x)$ is a crisp input of a fuzzy inference system with a residual Gödel rule base*

$$R(x, y) = \bigwedge_{i=1}^{n} A_i(x) \to_T B_i(y)$$

*then*

$$B'(y) = \bigwedge_{i=1}^{n} \alpha_i \to_T B_i(y),$$

*where $\alpha_i = A_i(x_0)$ is the firing strength of the $i^{th}$ fuzzy rule.*

**Proof.** The proof is left as an exercise.                                    ∎

These properties suggest a simplified implementation of a SISO fuzzy system, when the input of the system is crisp.

**Algorithm 7.7.**     *1. Input the crisp value $x_0$.*

   *2. Calculate the firing strengths of each fuzzy rule: $\alpha_i = A_i(x_0)$.*

   *3. Calculate for a given $y \in Y$ the output fuzzy set that is*

$$B'(y) = \bigvee_{i=1}^{n} \alpha_i \wedge B_i(y)$$

   *for a Mamdani rule base.*

   *4. Defuzzify $B'$ ($y_0 = defuzz(B')$.)*

   *5. Output $y_0$.*

**Example 7.8.** *Let us consider a simple Mamdani SISO fuzzy system with two fuzzy rules*

$$\text{If } x \text{ is } A_i \text{ then } y \text{ is } B_i, i = 1, 2,$$

*with $A_1 = (1, 2, 3)$, $A_2 = (2, 3, 4)$, $B_1 = (2, 4, 6)$, $B_2 = (4, 6, 8)$. Let $x_0 = 2.25$. Then the firing strength of the first rule is 0.75 while the firing strength of the second rule is 0.25. Then the output of the system is*

$$B'(y) = (0.75 \wedge B_1(y)) \vee (0.25 \wedge B_2(y)).$$

*The output can be expressed as*

$$B'(y) = \begin{cases} \frac{y-2}{2} & \text{if} \quad 2 \le y < 3.5 \\ 0.75 & \text{if} \quad 3.5 \le y < 4.5 \\ \frac{6-y}{2} & \text{if} \quad 4.5 \le y < 5.5 \\ 0.25 & \text{if} \quad 5.5 \le y < 7.5 \\ \frac{8-y}{2} & \text{if} \quad 7.5 \le y \le 8 \\ 0 & \text{otherwise} \end{cases}.$$

*Now we have to defuzzify this result. We have $\int_2^8 B'(y) \cdot y \cdot dy = 10.875$, $\int_2^8 B'(y) \cdot dy = 2.375$ and $COG(B') = 4.579$.*

**Remark 7.9.** *If the fuzzy rule base has more than one antecedent, similar conclusions hold true. Indeed, if e.g.*

$$R(x, y, z) = \bigvee_{i=1}^{n} A_i(x) \wedge B_i(y) \wedge C_i(z)$$

*is a Mamdani rule base in a Mamdani inference with crisp input $(x_0, y_0)$, then similar to the previous results we have*

$$C'(z) = \bigvee_{i=1}^{n} A_i(x_0) \wedge B_i(y_0) \wedge C_i(z)$$

$$= \bigvee_{i=1}^{n} \alpha_i \wedge C_i(z),$$

*where the firing strength of the $i^{th}$ rule is*

$$\alpha_i = A_i(x_0) \wedge B_i(y_0).$$

## 7.3 Approximation Properties of SISO Fuzzy Systems

Let us consider the problem of approximating an unknown continuous function $f : [a, b] \rightarrow [c, d]$, with $[c, d] \subseteq \mathbb{R}$ by a SISO Fuzzy System. We consider $[c, d]$ to be the range of the function $f$, i.e., $f([a, b]) = [c, d]$. The problem has been solved for the first time in B. Kosko's works (see [97]). There were several results, most of them showing that under some conditions, fuzzy systems are able to approximate any continuous function with arbitrary accuracy. In the approach presented here, the proofs of the approximation properties are constructive and error estimates are also provided. This approach is published in the present work for the first time.

Let $a = x_0 < x_1 < ... < x_n < x_{n+1} = b$ be a partition of the input domain with norm

$$\delta = \sup_{i=1,...,n+1} \{x_i - x_{i-1}\}.$$

Let $f(x_i) = y_i, i = 0, ..., n + 1$ be sample values of the function. Let us construct a SISO fuzzy system that describes the functional relationship $f(x) = y, x \in [a, b]$.

We consider fuzzy systems with the support of each fuzzy set in the antecedent part, satisfying $A_i(x_i) = 1$ and

$$(A_i)_0 = [x_{i-1}, x_{i+1}], i = 1, ..., n.$$

Additionally we consider $A_i(x)$ to be continuous. The fuzzy set $A_i$ can be seen as a model for the expression $x$ is about $x_i$, $i = 1, ..., n$.

The consequence part $B_i$ will be considered integrable and it is constructed such that $B_i(y_i) = 1$ and it has the support

$$(B_i)_0 = [\min\{y_{i-1}, y_i, y_{i+1}\}, \max\{y_{i-1}, y_i, y_{i+1}\}], i = 1, ..., n.$$

The fuzzy set $B_i$ can be seen as a model for the expression $y$ is about $y_i$, $i = 1, ..., n$.

The fuzzy rule base in this case becomes

If $x$ is about $x_i$ then $y$ is about $y_i, i = 1, ..., n$

or

If $x$ is $A_i$ then $y$ is $B_i, i = 1, ..., n$,

rule which is consistent with our intuitive idea on function approximation.

## 7.3.1    Approximation by Mamdani, Larsen and t-Norm Based SISO Fuzzy Systems

Now we consider a SISO fuzzy system with Mamdani rule base and Center of Gravity defuzzification. Let $x \in [a, b]$ be a crisp input. The fuzzy output is calculated as

$$B'(y) = \bigvee_{i=1}^{n} A_i(x) \wedge B_i(y),$$

which is subject to defuzzification and the result is

$$COG(B') = \frac{\int_c^d B'(y) \cdot y \cdot dy}{\int_c^d B'(y) dy}.$$

Combining these two relations we can write a SISO fuzzy system as

$$F(f, x) = \frac{\int_c^d \bigvee_{i=1}^{n}(A_i(x) \wedge B_i(y)) \cdot y \cdot dy}{\int_c^d \bigvee_{i=1}^{n}(A_i(x) \wedge B_i(y)) \cdot dy}.$$

Since $A_i$ is continuous, and $B_i$ integrable, $i = 1, ..., n$ we have $F(f, x)$ well defined and continuous.

Usually, the error estimates in classical approximation theory are provided in terms of the modulus of continuity (see appendix).

We will need two lemmas.

**Lemma 7.10.** *(see e.g. Bede-Nobuhara-Dankova-Di Nola)For any $a_i, b_i \in [0, \infty)$, $i \in \{0, ..., n\}$ we have*

$$\left| \bigvee_{i=0}^{n} a_i - \bigvee_{i=0}^{n} b_i \right| \leq \bigvee_{i=0}^{n} |a_i - b_i|.$$

**Proof.** Since max operation is non-decreasing in its arguments, we have

$$\bigvee_{i=0}^{n} a_i = \bigvee_{i=0}^{n} |b_i + a_i - b_i| \leq \bigvee_{i=0}^{n} b_i + |a_i - b_i|$$

and it follows

$$\bigvee_{i=0}^{n} a_i \leq \bigvee_{i=0}^{n} b_i + \bigvee_{i=0}^{n} |a_i - b_i|.$$

This inequality together with the symmetric case implies the statement of the lemma. ∎

**Remark 7.11.** *If $A_i, i = 1, ..., n$ are continuous and $B_i, i = 1, ..., n$ are integrable functions then taking into account that the max and min are continuous bi-variate functions, we obtain that the function $F(f, \cdot) : [a, b] \to [c, d]$ is continuous, being a composition of continuous functions. If we consider the operator $F(\cdot, x) : C[a, b] \to C[a, b]$ we observe that it is a nonlinear operator, so approximation by $F(f, x)$ can be studied in the framework of nonlinear approximation.*

The following approximation theorem can be obtained.

**Theorem 7.12.** *(Approximation property of the Mamdani SISO fuzzy system) Any continuous function $f : [a, b] \to \mathbb{R}$ can be uniformly approximated by a Mamdani SISO fuzzy system*

$$F(f, x) = \frac{\int_c^d [\bigvee_{i=1}^{n} A_i(x) \wedge B_i(y)] \cdot y \cdot dy}{\int_c^d \bigvee_{i=1}^{n} (A_i(x) \wedge B_i(y)) \cdot dy}$$

*with any membership functions for the antecedents and consequences $A_i$, $B_i, i = 1, ..., n$ satisfying*
*(i) $A_i$ continuous with $(A_i)_0 = [x_{i-1}, x_{i+1}], i = 1, ..., n$;*
*(ii) $B_i$ integrable with*

$$(B_i)_0 = [\min\{y_{i-1}, y_i, y_{i+1}\}, \max\{y_{i-1}, y_i, y_{i+1}\}]$$

*for $i = 1, ..., n$, where $y_i = f(x_i), i = 0, ..., n + 1$.*
*Moreover the following error estimate holds true*

$$\|F(f, x) - f(x)\| \leq 3\omega(f, \delta),$$

*with $\delta = \max_{i=1,...,n}\{x_i - x_{i-1}\}$.*

**Proof.** We observe that

$$\frac{\int_c^d [\bigvee_{i=1}^{n} A_i(x) \wedge B_i(y)] \cdot f(x) \cdot dy}{\int_c^d \bigvee_{i=1}^{n} (A_i(x) \wedge B_i(y)) \cdot dy} = f(x)$$

and we have

$$|F(f,x) - f(x)| = \left| \frac{\int_c^d [\bigvee_{i=1}^n (A_i(x) \wedge B_i(y))] \cdot (y - f(x)) \cdot dy}{\int_c^d \bigvee_{i=1}^n (A_i(x) \wedge B_i(y)) \cdot dy} \right|$$

$$\leq \frac{\int_c^d |[\bigvee_{i=1}^n (A_i(x) \wedge B_i(y))] \cdot (y - f(x))| \, dy}{\int_c^d \bigvee_{i=1}^n (A_i(x) \wedge B_i(y)) \cdot dy}.$$

By Lemma 7.10 we have

$$|F(f,x) - f(x)| \leq \frac{\int_c^d [\bigvee_{i=1}^n (A_i(x) \wedge B_i(y))] \cdot |y - f(x)| \, dy}{\int_c^d \bigvee_{i=1}^n (A_i(x) \wedge B_i(y)) \cdot dy}.$$

The membership degrees $A_i(x)$ are null outside of their support. Also, any fixed value of $x \in [a,b]$ belongs to the support of at most two fuzzy sets $A_i$. Suppose that $x \in (A_j)_0 \cup (A_{j+1})_0$. Then we have

$$|F(f,x) - f(x)|$$

$$\leq \frac{\int_c^d [(A_j(x) \wedge B_j(y)) \vee (A_{j+1}(x) \wedge B_{j+1}(y))] \cdot |y - f(x)| \, dy}{\int_c^d [(A_j(x) \wedge B_j(y)) \vee (A_{j+1}(x) \wedge B_{j+1}(y))] \cdot dy}$$

Taking now into account that

$$(B_i)_0 = [\min\{y_{i-1}, y_i, y_{i+1}\}, \max\{y_{i-1}, y_i, y_{i+1}\}]$$

we can restrict the integrals to the union of the supports of $B_j$ and $B_{j+1}$, which is a subset of

$$[\min\{y_{j-1}, y_j, y_{j+1}, y_{j+2}\}, \max\{y_{j-1}, y_j, y_{j+1}, y_{j+2}\}].$$

We obtain

$$|F(f,x) - f(x)|$$

$$\leq \frac{\int_{(B_j)_0 \cup (B_{j+1})_0} [(A_j(x) \wedge B_j(y)) \vee (A_{j+1}(x) \wedge B_{j+1}(y))] \cdot |y - f(x)| \, dy}{\int_{(B_j)_0 \cup (B_{j+1})_0} [(A_j(x) \wedge B_j(y)) \vee (A_{j+1}(x) \wedge B_{j+1}(y))] \cdot dy}.$$

By the definition of $B_j$, $B_{j+1}$, and by the intermediate value theorem applied to the continuous function $f$, given $y \in (B_j)_0 \cup (B_{j+1})_0$ there exist $z \in [x_{j-1}, x_{j+2}]$ such that $f(z) = y$. Then using the properties of the modulus of continuity in Theorem A.39 we get

$$|y - f(x)| = |f(z) - f(x)| \leq \omega(f, |x_{j+2} - x_{j-1}|)$$

$$\leq \omega(f, 3\delta) \leq 3\omega(f, \delta),$$

with $\delta = \max_{i=1,\dots,n}\{x_i - x_{i-1}\}$. Finally we obtain

$$|F(f,x) - f(x)| \leq 3\omega(f, \delta).$$

∎

**Corollary 7.13.** *A Mamdani SISO fuzzy system $F(f, x)$ is able to approximate any continuous function $f(x)$ with arbitrary accuracy.*

**Proof.** From Theorem A.39 we have If $\delta \to 0$ then $F(f, x) \to f(x)$ and the result is immediate. ∎

**Corollary 7.14.** *If we consider $f : [0, 1] \to [c, d]$ a continuous function and the partition $a = x_0 < x_1 < \ldots < x_{n+1} = b$ with equally spaced knots, then we have*

$$\|F(f, x) - f(x)\| \leq 3\omega \left( f, \frac{1}{n+1} \right).$$

**Proof.** From Theorem 7.12 the result is immediate. ∎

**Remark 7.15.** *We observe that the fuzzy sets $A_i, B_i$ do not need to be normal. A small support ensures the convergence of a fuzzy system to the target function. Also, we can consider fuzzy systems with more than two overlapping supports at a time. For example if $(A_j)_0, (A_{j+1})_0, \ldots, (A_{j+r})_0$ overlap and likewise $(B_j)_0, (B_{j+1})_0, \ldots, (B_{j+r})_0$, then one can obtain the error estimate*

$$\|F(f, x) - f(x)\| \leq (r + 1)\omega(f, \delta)$$

*in a similar way to Theorem 7.12.*

The preceding Theorem can be extended to Mamdani Fuzzy systems with Larsen rule base. In fact one can extend the Theorem to any t-norm.

**Theorem 7.16.** *(Approximation property of the Larsen SISO fuzzy system) Any continuous function $f : [a, b] \to \mathbb{R}$ can be uniformly approximated by a Mamdani SISO fuzzy system with Larsen rule base*

$$F(f, x) = \frac{\int_c^d [\bigvee_{i=1}^n A_i(x) \cdot B_i(y)] \cdot y \cdot dy}{\int_c^d [\bigvee_{i=1}^n A_i(x) \cdot B_i(y)] \cdot dy}$$

*with any membership functions for the antecedents and consequences $A_i$, $B_i, i = 1, \ldots, n$ satisfying*
*(i) $A_i$ continuous, $(A_i)_0 = [x_{i-1}, x_{i+1}], i = 1, \ldots, n$;*
*(ii) $B_i$ integrable,*

$$(B_i)_0 = [\min\{y_{i-1}, y_i, y_{i+1}\}, \max\{y_{i-1}, y_i, y_{i+1}\}],$$

$y_i = f(x_i), i = 1, \ldots, n$.
*Moreover the following error estimate holds true*

$$\|F(f, x) - f(x)\| \leq 3\omega(f, \delta),$$

*with $\delta = \max_{i=1,\ldots,n}\{x_i - x_{i-1}\}$.*

**Proof.** The proof is left as an exercise. ∎

**Corollary 7.17.** *A Larsen SISO fuzzy system $F(f,x)$ is able to approximate any continuous function $f(x)$ with arbitrary accuracy.*

**Proof.** The proof is immediate.                                                                                    ∎

**Theorem 7.18.** *(Approximation property of the t-norm based SISO fuzzy system) Any continuous function $f : [a,b] \to \mathbb{R}$ can be uniformly approximated by a Mamdani SISO fuzzy system with t-norm (we consider a continuous t-norm $T$) rule base*

$$F(f,x) = \frac{\int_c^d [\bigvee_{i=1}^n A_i(x) T B_i(y)] \cdot y \cdot dy}{\int_c^d [\bigvee_{i=1}^n A_i(x) T B_i(y)] \cdot dy}$$

*with any membership functions for the antecedents and consequences $A_i$, $B_i, i = 1, ..., n$ satisfying*
  *(i) $A_i$ continuous, $(A_i)_0 = [x_{i-1}, x_{i+1}], i = 1, ..., n;$*
  *(ii) $B_i$ integrable,*

$$(B_i)_0 = [\min\{y_{i-1}, y_i, y_{i+1}\}, \max\{y_{i-1}, y_i, y_{i+1}\}],$$

$y_i = f(x_i), i = 1, ..., n.$
*Moreover the following error estimate holds true*

$$\|F(f,x) - f(x)\| \le 3\omega\,(f,\delta)\,,$$

*with $\delta = \max_{i=1,...,n}\{x_i - x_{i-1}\}$.*

**Proof.** The proof is left as an exercise for the reader.                           ∎

## 7.3.2    Approximation by SISO Fuzzy System of Gödel and Gödel Residual Types

Similar to the Mamdani case we consider a continuous function $f : [a,b] \to [c,d]$, and $a = x_0 < x_1 < ... < x_n < x_{n+1} = b$ be a partition of the input domain with norm

$$\delta = \sup_{i=1,...,n+1} \{x_i - x_{i-1}\}.$$

Let $f(x_i) = y_i, i = 0, ..., n+1$ be known values of the function. The antecedents satisfy $A_i(x_i) = 1$ and

$$(A_i)_0 = [x_{i-1}, x_{i+1}], i = 1, ..., n.$$

Additionally we consider $A_i$ to be continuous.
   The consequences $B_i$ are such that $B_i(y_i) = 1$ and

$$(B_i)_0 = [\min\{y_{i-1}, y_i, y_{i+1}\}, \max\{y_{i-1}, y_i, y_{i+1}\}], i = 1, ..., n.$$

and they are supposed to be integrable. Now we consider a SISO fuzzy system with Gödel rule base and Center of Gravity defuzzification. Let $x \in [a, b]$ be a crisp input. The fuzzy output is calculated as

$$B'(y) = \bigwedge_{i=1}^{n} A_i(x) \to B_i(y),$$

which is subject to defuzzification and the result is

$$COG(B') = \frac{\int_c^d B'(y) \cdot y \cdot dy}{\int_c^d B'(y) dy}.$$

The continuity of $A_i$ and the integrability of $B_i$ ensures the integrability of $B'$.

Combining these two relations we can write a SISO fuzzy system as

$$F(f, x) = \frac{\int_c^d [\bigwedge_{i=1}^{n} A_i(x) \to B_i(y)] \cdot y \cdot dy}{\int_c^d [\bigwedge_{i=1}^{n} A_i(x) \to B_i(y)] \cdot dy}.$$

Let us remark here that the implication $\to$ is not continuous as a bi-variate function but since $B_i$ are integrable the output $F(f, x)$ is well defined. In general we cannot conclude that $F(f, x)$ is continuous.

We will need the following Lemma.

**Lemma 7.19.** *For any $a_i \in [0, \infty)$, $i \in \{0, ..., n\}$ and $y, z \in [0, \infty)$ we have*

$$\left| \bigwedge_{i=0}^{n} a_i y - \bigwedge_{i=0}^{n} a_i z \right| \leq \bigwedge_{i=0}^{n} a_i \cdot |y - z|.$$

**Proof.** We observe that

$$\bigwedge_{i=0}^{n} a_i y = \bigwedge_{i=0}^{n} a_i(z + y - z) \leq \bigwedge_{i=0}^{n} a_i(z + |y - z|)$$

$$\leq \bigwedge_{i=0}^{n} a_i z + \bigwedge_{i=0}^{n} a_i |y - z|$$

and symmetrically

$$\bigwedge_{i=0}^{n} a_i z \leq \bigwedge_{i=0}^{n} a_i y + \bigwedge_{i=0}^{n} a_i |y - z|.$$

The required inequality follows by combining the two cases.                    ∎

**Remark 7.20.** *A result similar to Lemma 7.10 does not hold. In fact, the following inequality can be proved*

$$\left| \bigwedge_{i=0}^{n} a_i - \bigwedge_{i=0}^{n} b_i \right| \leq \bigvee_{i=0}^{n} |a_i - b_i|,$$

*but this inequality cannot be used in the approximation result we are about to prove. This inconvenience was resolved by Lemma 7.19.*

**Theorem 7.21.** *(Approximation property of the Gödel SISO fuzzy system) Any continuous function $f : [a, b] \to \mathbb{R}$ can be uniformly approximated by a Gödel SISO fuzzy system*

$$F(f, x) = \frac{\int_c^d \left[ \bigwedge_{i=1}^{n} A_i(x) \to B_i(y) \right] \cdot y \cdot dy}{\int_c^d \left[ \bigwedge_{i=1}^{n} A_i(x) \to B_i(y) \right] \cdot dy}$$

*with any membership functions for the antecedents and consequences $A_i$, $B_i$, $i = 1, ..., n$ satisfying*
  *(i) $A_i$ continuous $(A_i)_0 = [x_{i-1}, x_{i+1}]$, $i = 1, ..., n$;*
  *(ii) $B_i$ are integrable*

$$(B_i)_0 = [\min\{y_{i-1}, y_i, y_{i+1}\}, \max\{y_{i-1}, y_i, y_{i+1}\}]$$

*for $i = 1, ..., n$, where $y_i = f(x_i)$, $i = 0, ..., n+1$.*
  *Moreover the following error estimate holds true*

$$\|F(f, x) - f(x)\| \leq 3\omega(f, \delta),$$

*with $\delta = \max_{i=1,...,n}\{x_i - x_{i-1}\}$.*

**Proof.** We observe that

$$\frac{\int_c^d \left[ \bigwedge_{i=1}^{n} A_i(x) \to B_i(y) \right] \cdot f(x) \cdot dy}{\int_c^d \left[ \bigwedge_{i=1}^{n} A_i(x) \to B_i(y) \right] \cdot dy} = f(x)$$

and we have

$$|F(f, x) - f(x)| \leq \frac{\int_c^d \left| \left[ \bigwedge_{i=1}^{n} A_i(x) \to B_i(y) \right] \cdot (y - f(x)) \right| dy}{\int_c^d \left[ \bigwedge_{i=1}^{n} A_i(x) \to B_i(y) \right] \cdot dy}.$$

By Lemma 7.19 we obtain

$$|F(f, x) - f(x)| \leq \frac{\int_c^d \left[ \bigwedge_{i=1}^{n} A_i(x) \to B_i(y) \right] \cdot |y - f(x)| \, dy}{\int_c^d \left[ \bigwedge_{i=1}^{n} A_i(x) \to B_i(y) \right] \cdot dy}.$$

The membership degrees $A_i(x)$ are null outside of their support so the implication $A_i(x) \to B_i(y) = 1$ whenever $x \notin (A_i)_0$. Also, any fixed value of

$x \in [a, b]$ belongs to the support of at most two fuzzy sets $A_i$. Suppose that $x \in (A_j)_0 \cup (A_{j+1})_0$, then we have

$$|F(f, x) - f(x)|$$

$$\leq \frac{\int_c^d [(A_j(x) \to B_j(y)) \wedge (A_{j+1}(x) \to B_{j+1}(y))] \cdot |y - f(x)| \, dy}{\int_c^d [(A_j(x) \to B_j(y)) \wedge (A_{j+1}(x) \to B_{j+1}(y))] \cdot dy}.$$

Taking now into account that

$$(B_i)_0 = [\min\{y_{i-1}, y_i, y_{i+1}\}, \max\{y_{i-1}, y_i, y_{i+1}\}]$$

we can restrict the integrals to the union of the supports of $B_j$ and $B_{j+1}$, and by the intermediate value theorem given $y \in (B_j)_0 \cup (B_{j+1})_0$ there exist $z \in [x_{j-1}, x_{j+2}]$ such that $f(z) = y$. Then using Theorem A.39 we obtain

$$|y - f(x)| \leq 3\omega(f, \delta)$$

with $\delta = \max_{i=1,\ldots,n}\{x_i - x_{i-1}\}$. Finally we obtain

$$|F(f, x) - f(x)| \leq 3\omega(f, \delta).$$

∎

**Corollary 7.22.** *A Gödel SISO fuzzy system $F(f, x)$ is able to approximate any continuous function $f(x)$ with arbitrary accuracy.*

**Proof.** The proof is immediate. ∎

For a residual rule based system we obtain the following property.

**Theorem 7.23.** *(Approximation property of the residual SISO fuzzy system) Any continuous function $f : [a, b] \to \mathbb{R}$ can be uniformly approximated by a Gödel residual SISO fuzzy system*

$$F(f, x) = \frac{\int_c^d [\bigwedge_{i=1}^n A_i(x) \to_T B_i(y)] \cdot y \cdot dy}{\int_c^d [\bigwedge_{i=1}^n A_i(x) \to_T B_i(y)] \cdot dy}$$

*with any membership functions for the antecedents and consequences $A_i, B_i$, $i = 1, \ldots, n$ satisfying*
*(i) $A_i$ are continuous and $(A_i)_0 = [x_{i-1}, x_{i+1}]$, $i = 1, \ldots, n$;*
*(ii) $B_i$ are integrable with*

$$(B_i)_0 = [\min\{y_{i-1}, y_i, y_{i+1}\}, \max\{y_{i-1}, y_i, y_{i+1}\}]$$

*for $i = 1, \ldots, n$, where $y_i = f(x_i)$, $i = 0, \ldots, n + 1$.*
*Moreover the following error estimate holds true*

$$\|F(f, x) - f(x)\| \leq 3\omega(f, \delta),$$

*with $\delta = \max_{i=1,\ldots,n}\{x_i - x_{i-1}\}$.*

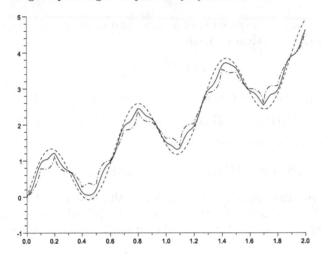

**Fig. 7.2** Approximation by Mamdani Fuzzy System. The function $f(x)$ (dashed) is approximated by Mamdani SISO fuzzy system with triangular (dash-dot) and Gaussian (solid line) membership functions.

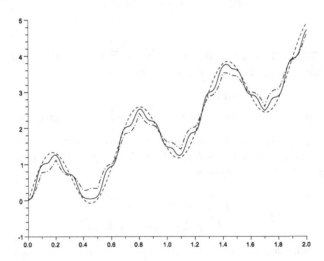

**Fig. 7.3** Approximation by Larsen Fuzzy System. The function $f(x)$ (dashed) is approximated by Larsen SISO fuzzy system with triangular (dash-dot) and Gaussian (solid line) membership functions.

**Proof.** The proof is left as an exercise.                                    ∎

**Corollary 7.24.** *A residual SISO fuzzy system $F(f, x)$ is able to approximate any continuous function $f(x)$ with arbitrary accuracy.*

**Proof.** The proof is immediate.                                              ∎

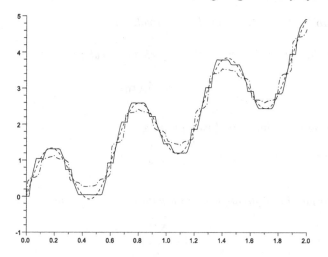

**Fig. 7.4** Approximation by Gödel Fuzzy System. The function $f(x)$ (dashed) is approximated by Gödel SISO fuzzy system with triangular (dash-dot) and Gaussian (solid line) membership functions.

**Example 7.25.** *We consider the function* $f : [0,2] \to \mathbb{R}$, $f(x) = 2x + \sin 10x$. *The function is approximated by Mamdani (Fig. 7.2) Larsen (Fig. 7.3) and Gödel (Fig. 7.4) SISO fuzzy systems. The membership functions of antecedents and consequences are of triangular and Gaussian types and there are 20 fuzzy rules describing each system.*

## 7.4   Takagi-Sugeno Fuzzy System

**Takagi-Sugeno fuzzy systems** (Sugeno [138]) are intrinsically single input single output systems. A Takagi-Sugeno system has the rule base consisting of antecedents of linguistic type and conclusions that are piecewise linear crisp outputs. This makes the defuzzification step redundant.

$$premise : \text{ If } x \text{ is } A_i \text{ then } z_i = a_i x + b_i, \ i = 1, ..., n$$

$$fact : \ x \text{ is } x_0$$

$$conclusion : \ z \text{ is } z_0.$$

Takagi-Sugeno (TS) fuzzy controllers do not use an inference system as the Mamdani or Gödel type systems described in the previous section, instead they use the firing strength of each fuzzy rule in the computation of the conclusions. TS fuzzy controller has crisp inputs, singleton fuzzifier and practically it does not have a defuzzifier.

The control algorithm for a TS fuzzy system is as follows:

**Algorithm 7.26.**    *1. Input the crisp value $x_0$.*

*2. Calculate the firing strengths of each fuzzy rule:*

$$\alpha_i = A_i(x_0).$$

*3. Calculate the individual rule outputs*

$$z_i = a_i x_0 + b_i.$$

*4. Aggregate the individual rule outputs and the system's output is*

$$z_0 = \frac{\sum_{i=1}^{n} \alpha_i z_i}{\sum_{i=1}^{n} \alpha_i}.$$

A Takagi-Sugeno system with more antecedents can be described as follows:

*premise* : if $x$ is $A_i$ and $y$ is $B_i$ then $z_i = a_i x + b_i y + c_i$, $i = 1, ..., n$

*fact* : $x$ is $x_0$ and $y$ is $y_0$

*conclusion* : $z$ is $z_0$

The control algorithm for a TS fuzzy system of this type is modified as follows

**Algorithm 7.27.**    *1. Input the crisp values $x_0, y_0$.*

*2. Calculate the firing strengths of each fuzzy rule*

$$\alpha_i = A_i(x_0) \wedge B_i(y_0).$$

*3. Calculate the rule outputs*

$$z_i = a_i x_0 + b_i y_0 + c_i.$$

*4. The output is*

$$z_0 = \frac{\sum_{i=1}^{n} \alpha_i z_i}{\sum_{i=1}^{n} \alpha_i}.$$

**Remark 7.28.** *Designing the rule base for a TS fuzzy system requires knowledge of the parameters $a_i, b_i, c_i$, $i = 1, ..., n$. The values of the parameters $a_i, b_i$ and $c_i$ can be given in advance or they can be obtained using adaptive techniques that will be discussed later.*

**Example 7.29.** *Consider the following TS fuzzy system*

$$\text{if } x \text{ is big and } y \text{ is small then } z_1 = x + y$$

$$\text{if } x \text{ is medium and } y \text{ is big then } z_2 = 2x - y.$$

*Consider the crisp inputs $x_0 = 3$ and $y_0 = 2$. Also, we know that $\mu_{big}(3) = 0.8$, $\mu_{small}(2) = 0.2$, $\mu_{medium}(3) = 0.6$, $\mu_{big}(2) = 0.9$.*
    *We can compute the firing strength of the rules*

$$\alpha_1 = 0.8 \wedge 0.2 = 0.2, \quad \alpha_2 = 0.6 \wedge 0.9 = 0.6$$

*The rule outputs are*

$$z_1 = x_0 + y_0 = 5, \quad z_2 = 2x_0 - y_0 = 4$$

*and the output of the system is*

$$z_0 = \frac{0.2 \cdot 5 + 0.6 \cdot 4}{0.2 + 0.6} = 4.25.$$

## 7.5   Approximation Properties of Takagi-Sugeno Fuzzy Systems

Let us consider the problem of approximating an unknown continuous function $f : [a, b] \to [c, d]$, with $[c, d] \subseteq \mathbb{R}$ by a Takagi-Sugeno Fuzzy System. In the present section we adapt a constructive approach. Let $a = x_0 \leq x_1 \leq \dots \leq x_n \leq x_{n+1} = b$ be a partition of the input domain and let $f(x_i) = y_i$, $i = 1, \dots, n$, be sample data. Let $A_i \in \mathcal{F}([a, b]), i = 1, \dots, n$ be fuzzy sets with continuous membership degree such that $(A_i)_0 = [x_{i-1}, x_{i+1}]$. The output of the Takagi-Sugeno fuzzy System can be written as

$$F(f, x) = \frac{\sum_{i=1}^{n} A_i(x) \cdot (a_i x + b_i)}{\sum_{i=1}^{n} A_i(x)},$$

with $a_i, b_i \in \mathbb{R}$.

Also, it is possible to generalize a TS fuzzy system such that the individual rule outputs are polynomials.

$$F(f, x) = \frac{\sum_{i=1}^{n} A_i(x) \cdot (a_{m,i} x^m + a_{m-1,i} x^{m-1} + \dots + a_{1,i} x + a_{0,i})}{\sum_{i=1}^{n} A_i(x)}.$$

This combination builds a higher-order Takagi-Sugeno fuzzy system. The idea to consider combinations between different approximation operators is not new, it was used before in several works in Approximation Theory see Gal [67], open problem 1.6.11. Also in the same direction see e.g. Lucyna

[102], where the authors combine approximation operators of Favard type with Taylor series and Coman [39], for a Shepard-Taylor combined operator (see also Stancu-Coman-Blaga [134]). These ideas can be used successfully to prove that Takagi-Sugeno fuzzy systems have very good approximation properties.

In a Takagi-Sugeno fuzzy system the coefficients $a_i, b_i, i = 1, .., n$ (for the linear individual rule output) and respectively $a_{k,i}, i = 1, ..., n, k = 0, ..., m$ (for higher order Takagi-Sugeno system) are obtained through an optimization process.

We adopt a constructive approach here. Surely, starting with the values that we select further optimization is still possible.

**Theorem 7.30.** *(Approximation property of Takagi-Sugeno fuzzy system) Any continuously differentiable function $f : [a, b] \to \mathbb{R}$ can be uniformly approximated by a Takagi-Sugeno fuzzy system defined as*

$$F(f, x) = \frac{\sum_{i=1}^n A_i(x) \cdot (a_i(x - x_i) + b_i)}{\sum_{i=1}^n A_i(x)},$$

*with the antecedents $A_i, i = 1, ..., n$ continuous, satisfying $(A_i)_0 = [x_{i-1}, x_{i+1}]$, $i = 1, ..., n$ and the individual rule outputs satisfying $b_i = f(x_i)$ and $a_i = f'(x_i)$, $i = 1, ..., n$.*

*Moreover the following error estimate holds true*

$$\|F(f, x) - f(x)\| \leq 9\delta\omega\left(f', \delta\right),$$

*with $\delta = \max_{i=1,...,n}\{x_i - x_{i-1}\}$.*

**Proof.** Let us start with Lagrange's Theorem. If $f$ is continuous on $[x_i, x]$ (or $[x, x_i]$ if $x < x_i$) and differentiable on $(x_i, x)$ (or $(x, x_i)$) then there is a $c_i$ between $x$ and $x_i$ such that

$$f(x) = f'(c_i)(x - x_i) + f(x_i).$$

Then we have

$$f(x) = \frac{\sum_{i=1}^n A_i(x) \cdot f(x)}{\sum_{i=1}^n A_i(x)}$$

$$= \frac{\sum_{i=1}^n A_i(x) \cdot (f'(c_i)(x - x_i) + f(x_i))}{\sum_{i=1}^n A_i(x)}.$$

The Takagi-Sugeno fuzzy system can be written

$$F(f, x) = \frac{\sum_{i=1}^n A_i(x) \cdot (f'(x_i)(x - x_i) + f(x_i))}{\sum_{i=1}^n A_i(x)}.$$

Then we have

$$|F(f, x) - f(x)| = \left| \frac{\sum_{i=1}^n A_i(x) \cdot (f'(c_i) - f'(x_i))(x - x_i)}{\sum_{i=1}^n A_i(x)} \right|$$

$$\leq \frac{\sum_{i=1}^{n} A_i(x) \cdot |f'(c_i) - f'(x_i)| \, |x - x_i|}{\sum_{i=1}^{n} A_i(x)}.$$

For a fixed value of $x$ there are only two nonzero terms of the sum i.e., $x \in [x_{j-1}, x_{j+2}]$ produces the only two nonzero terms $A_j(x)$ and $A_{j+1}(x)$. Then, taking into account properties of the modulus of continuity we obtain

$$|F(f,x) - f(x)| \leq \frac{\sum_{i=1}^{n} A_i(x) \cdot \omega(f', |c_i - x_i|) \, |x - x_i|}{\sum_{i=1}^{n} A_i(x)}$$

$$\leq \frac{\sum_{i=1}^{n} A_i(x) \cdot \omega(f', 3\delta) 3\delta}{\sum_{i=1}^{n} A_i(x)} \leq 9\delta\omega(f', \delta).$$

∎

**Corollary 7.31.** *Let $f$ be twice continuously differentiable. Then the following quadratic error estimate holds true*

$$\|F(f,x) - f(x)\| \leq 9\delta^2 \, \|f''\| \, ,$$

*here the norm is the supremum norm.*

**Proof.** The proof is immediate.                    ∎

**Remark 7.32.** *We observe that the fuzzy sets $A_i$, do not need to be normal. Also, we can consider fuzzy systems with more than two overlapping supports at a time. For example if*

$$(A_j)_0, (A_{j+1})_0, ..., (A_{j+r})_0$$

*overlap, then one can obtain the error estimate*

$$\|F(f,x) - f(x)\| \leq (r+1)^2 \delta\omega(f', \delta).$$

**Remark 7.33.** *We observe that the Takagi-Sugeno fuzzy system constructed with fixed membership functions $A_i$ is a linear operator. So, TS fuzzy systems can be studied using techniques of classical approximation theory. But a TS fuzzy system combined with adaptive techniques will not be a linear operator any more.*

**Theorem 7.34.** *(Approximation property of higher-order Takagi-Sugeno fuzzy system) Let $f : [a,b] \to \mathbb{R}$ be $m$ times continuously differentiable. The it can be uniformly approximated by a higher-order Takagi-Sugeno fuzzy system defined as*

$$F(f,x) = \frac{\sum_{i=1}^{n} A_i(x) \cdot (a_{m,i}(x - x_i)^m + ... + a_{1,i}(x - x_i) + a_{0,i})}{\sum_{i=1}^{n} A_i(x)}.$$

*with the antecedents $A_i, i = 1, ..., n$ satisfying $(A_i)_0 = [x_{i-1}, x_{i+1}]$, $i = 1, ..., n$ and the individual rule outputs being Taylor polynomials about $x_i$ of $f(x)$, i.e.,*

$$a_{k,i} = \frac{f^{(k)}(x_i)}{k!}.$$

*Moreover the following error estimate holds true*

$$\|F(f,x) - f(x)\| \leq \frac{3^{m+1}}{m!} \delta^m \omega(f^{(m)}, \delta)$$

*with $\delta = \max_{i=1,\dots,n}\{x_i - x_{i-1}\}$.*

**Proof.** We start with a Taylor expansion which insures the existence of $\xi_i$ between $x$ and $x_i$ such that

$$f(x) = f(x_i) + \frac{f'(x_i)}{1!}(x - x_i) + \dots + \frac{f^{(m)}(\xi_i)}{m!}(x - x_i)^m.$$

Then we have

$$f(x) = \frac{\sum_{i=1}^{n} A_i(x) \cdot f(x)}{\sum_{i=1}^{n} A_i(x)}$$

and so

$$f(x) = \frac{\sum_{i=1}^{n} A_i(x) \cdot \left[f(x_i) + \frac{f'(x_i)}{1!}(x - x_i) + \dots + \frac{f^{(m)}(\xi_i)}{m!}(x - x_i)^m\right]}{\sum_{i=1}^{n} A_i(x)}.$$

The Takagi-Sugeno fuzzy system can be written

$$F(f,x) = \frac{\sum_{i=1}^{n} A_i(x) \cdot \left[f(x_i) + \frac{f'(x_i)}{1!}(x - x_i) + \dots + \frac{f^{(m)}(x_i)}{m!}(x - x_i)^m\right]}{\sum_{i=1}^{n} A_i(x)}.$$

Then we have

$$|F(f,x) - f(x)| = \frac{\sum_{i=1}^{n} A_i(x) \cdot \left|\frac{f^{(m)}(x_i)}{m!} - \frac{f^{(m)}(\xi_i)}{m!}\right| |x - x_i|^m}{\sum_{i=1}^{n} A_i(x)}.$$

Taking into account that $x \in [x_{j-1}, x_{j+2}]$ produces the only two nonzero terms $A_j(x)$ and $A_{j+1}(x)$, we obtain

$$|F(f,x) - f(x)| \leq \frac{\sum_{i=1}^{n} A_i(x) \cdot \frac{1}{m!}\omega(f^{(m)}, |x_i - \xi_i|) |x - x_i|^m}{\sum_{i=1}^{n} A_i(x)}$$

$$\leq \frac{\sum_{i=1}^{n} A_i(x) \cdot \frac{1}{m!}\omega(f^{(m)}, 3\delta)(3\delta)^m}{\sum_{i=1}^{n} A_i(x)}$$

$$\leq \frac{3^{m+1}}{m!} \delta^m \omega(f^{(m)}, \delta).$$

∎

**Corollary 7.35.** *Let $f$ be $m+1$ times continuously differentiable. Then the following error estimate holds true*

$$\|F(f,x) - f(x)\| \leq \frac{3^{m+1}}{(m+1)!} \delta^{m+1} \left\|f^{(m+1)}\right\|$$

*here the norm is the supremum norm.*

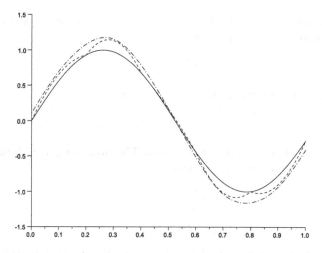

**Fig. 7.5** Approximation by Takagi-Sugeno Fuzzy System. $f(x) = sin(6x)$ (solid line) is approximated by Takagi-Sugeno system with triangular (dashed line) and Gaussian (dash-dot line) membership functions.

**Proof.** The proof is immediate. ∎

**Remark 7.36.** *We observe that the fuzzy sets $A_i$, do not need to be normal. Also, we can consider fuzzy systems with more than two overlapping supports in this higher order system as well. If $(A_j)_0, (A_{j+1})_0, ..., (A_{j+r})_0$ have non-empty intersection, then one can obtain the error estimate*

$$\|F(f,x) - f(x)\| \leq \frac{(r+1)^{m+1}}{(m+1)!} \delta^{m+1} \left\| f^{(m+1)} \right\|.$$

**Example 7.37.** *We consider the problem of approximating $f(x) = \sin(6x)$, $x \in [0,1]$ by a Takagi Sugeno fuzzy system with 5 rules.*

$$premise: \quad If \ x \ is \ A_i \ then \ z_i = a_i x + b_i, \ i = 1, ..., n$$

*We can construct the TS fuzzy system in Theorem 7.30 using Gaussian or triangular membership functions. The results can be compared in Fig. 7.5.*

## 7.6    Fuzzy Control

A dynamical system is a function which describes the time-dependence of a point's position in space

$$\Phi : U \subseteq T \times M \to M,$$

where $T$ is either a real interval (for continuous dynamical systems) or it is a subset of integers (for discrete dynamical systems), $M$ is a set of states, coordinates for the system. The behavior of a dynamical system is often described by differential equations

$$x' = f(t, x)$$

for continuous-time systems, or difference equations

$$x_{n+1} = f(t_n, x_n, ..., x_0),$$

for the case of discrete dynamical system. The function $f(t, x)$ besides time and coordinate may depend on parameters

$$x' = f(t, x, u).$$

$$x_{n+1} = f(t_n, x_n, ..., x_0, u).$$

This allows us to get a desired position or trajectory by manipulating parameters of the system. The evolution of the values of the parameter $u$ that give the desired position (track) is called the control signal. Classical control engineering uses mathematical analysis to update in time the control signal $u$. The approach needs intense computational and theoretical efforts.

The idea of Zadeh was to formulate the control algorithmically in terms of simple (fuzzy) logical rules. This idea lead to the development of fuzzy controllers.

A fuzzy controller aims to implement fuzzy inference systems in control theory. The first fuzzy controller was designed by Mamdani and Assilian [107]. Since then many applications have used fuzzy controllers and there are several types of fuzzy controllers. A fuzzy controller is essentially a SISO fuzzy inference system. Often one may make the confusion to call a SISO fuzzy inference system as a fuzzy controller. The usage of this terminology is inconsistent with control theory. Fuzzy controller should have the output linked into a control system as described above. So in the current work we make a distinction between SISO fuzzy system and fuzzy controller.


## 7.7    Example of a Fuzzy Controller

An inverted pendulum has its lower end fixed to a cart (see Fig. 7.6). A fuzzy controller has the goal to balance the inverted pendulum.

The differential equation that describes an inverted pendulum is

$$l\ddot{\theta} - g\sin\theta = \ddot{x}\cos\theta$$

where $\theta, x$ are as in the figure. The inverted pendulum system can be controlled by a fuzzy system described by a set of fuzzy rules. The antecedent variables considered in this problem are $\theta$ and its derivative $\dot{\theta}$. The control

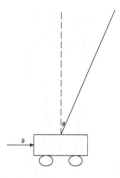

**Fig. 7.6** An inverted pendulum

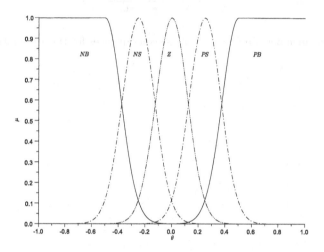

**Fig. 7.7** The membership functions of the antecedents for the angle $\theta$ and $\dot\theta$

signal is $u = \ddot{x}$, the acceleration of the cart. The antecedents we consider are positive big (PB), positive small (PS), near zero (Z), negative small (NS), and negative big (NB), both for $\theta$ and its derivative $\dot\theta$ and as well for the control signal $u$ (see Fig. 7.7).

All the fuzzy sets we consider are described by Gaussian membership functions with parameters $x_1, x_2, \sigma$ different for different variables.

$$PB(x) = \left\{ \begin{array}{ll} e^{-\frac{(x-x_1)^2}{2\sigma^2}} & \text{if} \quad x < x_1 \\ 1 & \text{if} \quad x_1 \leq x \end{array} \right.$$

$$PS(x) = e^{-\frac{(x-x_2)^2}{2\sigma^2}}, \ Z(x) = e^{-\frac{x^2}{2\sigma^2}}, \ NS(x) = e^{-\frac{(x+x_2)^2}{2\sigma^2}}$$

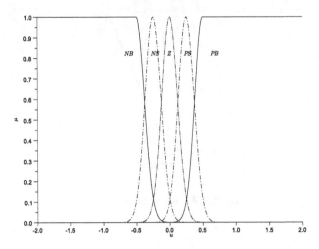

**Fig. 7.8** The membership functions of the consequences for the acceleration of the cart $\ddot{x}$

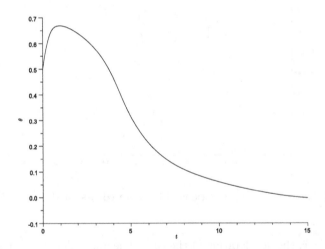

**Fig. 7.9** The time dependence of the angle $\theta$ for an inverted pendulum with a fuzzy controller

$$NB(x) = \begin{cases} e^{-\frac{(x+x_1)^2}{2\sigma^2}} & \text{if} \quad -x_1 < x \\ 1 & \text{if} \quad x \le -x_1 \end{cases}.$$

The fuzzy rule base consists of five fuzzy rules

If $\theta$ is PB and $\dot{\theta}$ is PB then $u$ is NB
If $\theta$ is PS and $\dot{\theta}$ is PS then $u$ is NS
If $\theta$ is Z and $\dot{\theta}$ is Z then $u$ is Z
If $\theta$ is NS and $\dot{\theta}$ is NS then $u$ is PS
If $\theta$ is NB and $\dot{\theta}$ is NB then $u$ is PB

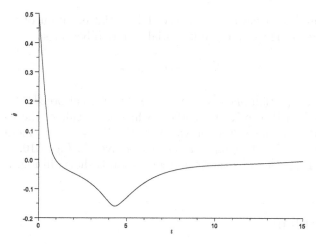

**Fig. 7.10** The time dependence of the angular velocity $\dot\theta$ for an inverted pendulum with a fuzzy controller

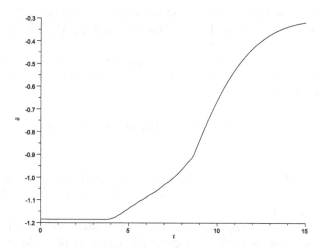

**Fig. 7.11** The control signal $\ddot{x}$ provided by a fuzzy controller

A Mamdani SISO fuzzy system provides the output $u(\theta, \dot\theta)$. The equation of the Mamdani system is

$$u(\theta, \dot\theta) = \frac{\int_c^d \left[ \bigvee_{i=1}^5 A_i(\theta) \wedge B_i(\dot\theta) \wedge C_i(z) \right] \cdot z \cdot dz}{\int_c^d \left[ \bigvee_{i=1}^5 A_i(\theta) \wedge B_i(\dot\theta) \wedge C_i(z) \right] \cdot dz},$$

and it is used as a parameter in the differential equation of the inverted pendulum to stabilize it. The differential equation becomes

$$l\ddot{\theta} - g\sin\theta = u\cos\theta.$$

To solve this equation one can use different numerical procedures. We have used a standard Runge-Kutta method, which is outside of the scope of the present text. The angle $\theta$ as provided by the Runge-Kutta method is represented in Fig. 7.9. The angular velocity is given in Fig. 7.10. The control signal that is provided by the Mamdani system is shown in Fig. 7.11.

## 7.8  Problems

1. Consider a triangular fuzzy number $(1, 2, 4)$. Find its center of gravity defuzzification. Find the expected value of the fuzzy number.

2. Compare the center of gravity and the center of area defuzzifications for fuzzy numbers of trapezoid type. (Numeric example $(1, 2, 3, 6)$).

3. Consider a fuzzy inference system with singleton fuzzifier and a given rule base $R(x, y)$. Prove that, for the output of an inference system with any t-norm we have

$$B'(y) = \bigvee_{x \in X} A'(x)TR(x, y) = R(x_0, y).$$

4. Consider a Mamdani fuzzy inference system with two antecedents $A_1 = (1, 3, 5)$ and $A_2 = (3, 5, 7)$ and two consequences $B_1 = (5, 10, 15)$ and $B_2 = (10, 15, 20)$.
   a) Let $x_0 = 4$ be a crisp input for the fuzzy system considered. Find the firing strength of each rule, then graph the output $B'$ of the fuzzy inference system.
   b) Let $x_0 = 3.5$ Find the firing strength of each rule, then graph the output $B'$ of the fuzzy inference system.
   c) Does the output change if we consider the same fuzzy rule base with a Gödel inference?

5. Design a SISO fuzzy system that controls a spaceship that has to avoid asteroids taking into account the closest one. The variables that can be considered are: asteroid is near, relatively close, far. The direction of the asteroid's velocity can be expressed as a linguistic variable as: approaching, not approaching, departing. Describe the fuzzy rules and fuzzy controller that could control the spaceship.

6. Consider the following TS fuzzy system

$$\text{if } x \text{ is } A_1 \text{ and } y \text{ is } B_1 \text{ then } z = 2x + y$$

$$\text{if } x \text{ is } A_2 \text{ and } y \text{ is } B_2 \text{ then } z = x - 2y$$

Consider the crisp inputs $x_0 = 5$ and $y_0 = 6$. Also, we know that $A_1(5) = 0.7$, $A_2(5) = 0.5$, $B_1(6) = 0.6$, $B_2(6) = 0.8$. Find the firing strengths of each rule and then find the output $z_0$ of the TS fuzzy controller.

7. Consider the fuzzy sets $A_1 = (1, 2, 3)$, $A_2 = (1, 3, 4)$, $B_1 = (10, 14, 18)$, and $B_2 = (14, 18, 20)$, and the following TS fuzzy system

$$\text{if } x \text{ is } A_1 \text{ and } y \text{ is } B_1 \text{ then } z = 5x + y$$

$$\text{if } x \text{ is } A_2 \text{ and } y \text{ is } B_2 \text{ then } z = 4x + y$$

Consider the crisp inputs $x_0 = 2$ and $y_0 = 16$. Find the firing strengths of each rule and then find the output $z_0$ of the TS fuzzy controller.

8. Describe and implement a SISO version of the room temperature control problem discussed in detail previously. The system has the input as the room temperature and the output is the defuzzification of one of the inference systems used (Mamdani, Larsen, t-norm, Gödel, Gödel residual).

9. We consider a ball and beam system. This consists of a ball that is rolling on a beam that can be tilted. The differential equation that describes a ball and beam system has the form

$$\ddot{x} = x(\dot{\theta})^2 - g \sin \theta$$

where $\theta$ is the angle between the beam and the horizontal direction, while $x$ is the coordinate of the ball. The ball and beam system can be controlled by a fuzzy system. The antecedent variables considered in this problem are $x$ and the angle $\theta$. The control signal is $u = \ddot{\theta}$, the angular acceleration of the system. The antecedents we consider are positive (P), and negative (N), both for $x$ and $\theta$. For the control signal $u$ we consider four fuzzy variables positive big (PB), positive small (PS), near zero (Z), negative small (NS), and negative big (NB). All the fuzzy sets we consider can have trapezoidal or Gaussian membership functions. The fuzzy sets PB, PS, Z, NS and NB are as in the inverted pendulum problem. The fuzzy rule base consists of five fuzzy rules of the form

If $x$ is P and $\theta$ is P then $u$ is NS

If $x$ is P and $\theta$ is N then $u$ is PB

If $x$ is Z and $\theta$ is Z then $u$ is Z

If $x$ is N and $\theta$ is P then $u$ is NB

If $x$ is N and $\theta$ is N then $u$ is NS

Construct and implement a fuzzy controller to perform the desired task.

# 8

# Fuzzy Analysis

The topological structure of Fuzzy Numbers was investigated in detail by several authors (e.g., Diamond-Kloeden[44], Puri-Ralescu [123], Ma [104], Goetschel-Voxman [74]). There are some properties that in a classical Mathematical structure (e.g. that of a Banach space) are easily fulfilled, while in the fuzzy setting they do not hold. In this sense, in this chapter we present several negative results through several counterexamples. Some of these results are known but the counterexamples presented in Sections 8.2, 8.3, 8.4 are new, being published for the first time here. Mathematical Analysis on Fuzzy Number's space is an interesting topic (see Anastassiou [4], Bede-Gal [18], [20], [19], Chalco-Cano-Román-Flores-Jiménez-Gamero [35] Gal [66], Lakshimikantham-Mohapatra [98] Wu-Gong [151]). We study in this chapter mainly integration and differentiability of fuzzy-number-valued functions.

## 8.1  Metric Spaces of Fuzzy Numbers

The most well known, and also the most employed metric in the space of fuzzy numbers is the Hausdorff distance. The Hausdorff (or in fact Hausdorff-Pompeiu) distance for fuzzy numbers (Erceg [56], Puri-Ralescu [123], Diamond-Kloeden[44]) is based on the classical Hausdorff-Pompeiu distance between compact convex subsets of $\mathbb{R}^n$. Let us recall the Definition of the Hausdorff distance in this case.

(i) Let $\mathcal{K}$ denote the collection of all nonempty compact convex subsets of $\mathbb{R}^n$ and let $A \in \mathcal{K}$. The distance from a point $x$ to the set $A$ is

B. Bede: *Mathematics of Fuzzy Sets and Fuzzy Logic*, STUDFUZZ 295, pp. 137–170.
DOI: 10.1007/978-3-642-35221-8_8     © Springer-Verlag Berlin Heidelberg 2013

$$d(x, A) = \inf\{||x - a|| : a \in A\}.$$

(ii) Let now $A, B \in \mathcal{K}$. The Hausdorff separation of $B$ from $A$ and $A$ from $B$ respectively, are

$$d_H^*(B, A) = \sup\{d(b, A) : b \in B\}$$
$$d_H^*(A, B) = \sup\{d(a, B) : a \in A\}.$$

(iii) The Hausdorff distance between $A, B \in \mathcal{K}$ is

$$d_H(A, B) = \max\{d_H^*(A, B), d_H^*(B, A)\}.$$

For the particular case when $A = [a_1, a_2]$, $B = [b_1, b_2]$ are two intervals (we denote $A, B \in \mathbb{I}$) the Hausdorff-Pompeiu distance is

$$d_H(A, B) = \max\{|a_1 - b_1|, |a_2 - b_2|\}.$$

It is known that with respect to the Hausdorff distance, $\mathcal{K}$ (and in particular $\mathbb{I}$) is a complete separable metric space. Let us now turn to our object that is the metric space of Fuzzy Numbers.

**Definition 8.1.** *(Diamond-Kloeden [44]) Let $D_\infty : \mathbb{R}_{\mathcal{F}} \times \mathbb{R}_{\mathcal{F}} \to \mathbb{R}_+ \bigcup \{0\}$,*

$$D_\infty(u, v) = \sup_{r \in [0,1]} \max\{|u_r^- - v_r^-|, |u_r^+ - v_r^+|\}$$

$$= \sup_{r \in [0,1]} \{d_H(u_r, v_r)\},$$

*where $u_r = [u_r^-, u_r^+]$, $v_r = [v_r^-, v_r^+] \subseteq \mathbb{R}$ and $d_H$ is the classical Hausdorff-Pompeiu distance between real intervals. Then $D_\infty$ is called the Hausdorff distance between fuzzy numbers.*

An $L^p$- type distance can also be defined

**Definition 8.2.** *(Diamond-Kloeden [44]) Let $1 \le p < \infty$. We define the $D_p$ distance between fuzzy numbers as*

$$D_p(u, v) = \left(\int_0^1 d_H(u_r, v_r)^p dr\right)^{1/p}$$

$$= \left(\int_0^1 \max\{|u_r^- - v_r^-|, |u_r^+ - v_r^+|\}^p dr\right)^{1/p}.$$

The next Theorem gives a list of properties of the Hausdorff distance between fuzzy numbers.

**Proposition 8.3.** *(Diamond-Kloeden [44]) (i)$(\mathbb{R}_{\mathcal{F}}, D_{\infty})$ is a metric space and moreover, the following properties hold true:*
(ii)
$$D_{\infty}\left(u + w, v + w\right) = D_{\infty}\left(u, v\right), \forall u, v, w \in \mathbb{R}_{\mathcal{F}},$$
*i.e., $D_{\infty}$ is translation invariant.*
(iii)
$$D_{\infty}\left(k \cdot u, k \cdot v\right) = |k|\, D_{\infty}\left(u, v\right), \forall u, v \in \mathbb{R}_{\mathcal{F}}, \ \forall k \in \mathbb{R};$$
(iv)
$$D_{\infty}\left(u + v, w + e\right) \leq D_{\infty}\left(u, w\right) + D_{\infty}\left(v, e\right), \forall u, v, w, e \in \mathbb{R}_{\mathcal{F}}.$$

**Proof.** (i) It is easy to see that
$$D_{\infty}\left(u, v\right) = \sup_{r \in [0,1]}\, \max\left\{\left|u_r^- - v_r^-\right|, \left|u_r^+ - v_r^+\right|\right\} \geq 0$$

with $D_{\infty}\left(u, v\right) = 0$ if and only if $u_r^- = v_r^-, u_r^+ = v_r^+, r \in [0, 1]$. Also it is obvious that $D_{\infty}\left(u, v\right) = D_{\infty}\left(v, u\right)$. For the triangle inequality we have
$$\left|u_r^- - w_r^-\right| \leq \left|u_r^- - v_r^-\right| + \left|v_r^- - w_r^-\right|$$
$$\leq D_{\infty}(u, v) + D_{\infty}(v, w)$$

and
$$\left|u_r^+ - w_r^+\right| \leq \left|u_r^+ - v_r^+\right| + \left|v_r^+ - w_r^+\right|$$
$$\leq D_{\infty}(u, v) + D_{\infty}(v, w)$$

which implies
$$D_{\infty}(u, w) \leq D_{\infty}(u, v) + D_{\infty}(v, w).$$

As a conclusion $(\mathbb{R}_{\mathcal{F}}, D_{\infty})$ is a metric space.
(ii) We have
$$D_{\infty}\left(u, v\right) = \sup_{r \in [0,1]}\, \max\left\{\left|u_r^- - v_r^-\right|, \left|u_r^+ - v_r^+\right|\right\}$$
$$= \sup_{r \in [0,1]}\, \max\left\{\left|u_r^- + w_r^- - v_r^- - w_r^-\right|, \left|u_r^+ + w_r^+ - v_r^+ - w_r^+\right|\right\}$$
$$= D_{\infty}(u + w, v + w).$$

(iii) It is easy to see that
$$D_{\infty}\left(ku, kv\right) = \sup_{r \in [0,1]}\, \max\left\{\left|ku_r^- - kv_r^-\right|, \left|ku_r^+ - kv_r^+\right|\right\}$$
$$= |k| D_{\infty}\left(u, v\right).$$

(iv) For (iv) we use the triangle inequality with the translation invariance. Indeed,
$$D_{\infty}(u + v, w + e) \leq D_{\infty}(u + v, w + v) + D_{\infty}(w + v, w + e)$$
$$= D_{\infty}(u, w) + D_{\infty}(v, e).$$

∎

**Proposition 8.4.** *(Diamond-Kloeden[44], Wu-Gong [151]) Let $1 \leq p < \infty$.*
*(i) Then $(\mathbb{R}_{\mathcal{F}}, D_p)$ is a metric space and moreover the following properties*
*hold true:*
   *(ii)*

$$D_p\left(u + w, v + w\right) = D_p\left(u, v\right), \forall u, v, w \in \mathbb{R}_{\mathcal{F}};$$

   *(iii)*

$$D_p\left(k \cdot u, k \cdot v\right) = |k| \, D_p\left(u, v\right), \forall u, v \in \mathbb{R}_{\mathcal{F}}, \ \forall k \in \mathbb{R};$$

   *(iv)*

$$D_p\left(u + v, w + e\right) \leq D_p\left(u, w\right) + D_p\left(v, e\right), \forall u, v, w, e \in \mathbb{R}_{\mathcal{F}}.$$

**Proof.** The proof is left to the reader as an exercise. $\blacksquare$

## 8.2  Completeness

The following Theorem shows that the Hausdorff distance defines a complete
metric on fuzzy numbers.

**Theorem 8.5.** *(Diamond-Kloeden[44]) $(\mathbb{R}_{\mathcal{F}}, D_\infty)$ is a complete metric space.*

**Proof.** For the proof, we need to show that any Cauchy sequence of fuzzy
numbers in $D_\infty$ is convergent in $\mathbb{R}_{\mathcal{F}}$. So let $u_n \in \mathbb{R}_{\mathcal{F}}, n \geq 1$ be a Cauchy
sequence. Then for any $\varepsilon > 0$ there exists $N \geq 1$ with

$$D_\infty(u_n, u_{n+p}) < \varepsilon, \forall n \geq N, p \geq 1,$$

i.e.,

$$\sup_{r \in [0,1]} \max \left\{ \left| (u_n)_r^- - (u_{n+p})_r^- \right|, \left| (u_n)_r^+ - (u_{n+p})_r^+ \right| \right\} < \varepsilon.$$

This last condition means that the sequences $(u_n)_r^-$ and $(u_n)_r^+$ are Cauchy
sequences in $\mathbb{R}$ so they converge as sequences of real numbers. Besides, $(u_n)_r$
also converges as a sequence of real intervals in the Hausdorff-Pompeiu distance. Let $u_r^-$ and $u_r^+$ be the limits of the sequences $(u_n)_r^-$ and $(u_n)_r^+$ respectively. It is easy to see that since $(u_n)_r^+ \leq (u_n)_r^+$ we will have $u_r^- \leq u_r^+$ and
$M_r = [u_r^-, u_r^+]$ is a well defined real interval. We will prove in the next step
that the sets $M_r$ satisfy the hypothesis of Negoita-Ralescu characterization
theorem 4.8.
   (i) The fact that $M_r$ is a real interval was proven above.
   (ii) Let $\alpha \leq \beta$. Then we have $(u_n)_\beta \subseteq (u_n)_\alpha$ i.e.,

$$(u_n)_\alpha^- \leq (u_n)_\beta^- \leq (u_n)_\beta^+ \leq (u_n)_\alpha^+$$

and by taking the limit in this inequality we obtain

$$u_\alpha^- \leq u_\beta^- \leq u_\beta^+ \leq u_\alpha^+,$$

which proves (ii).

(iii) Let $r_n$ be a convergent sequence $r_n \nearrow r$. By (ii) we have $u_r \subseteq u_{r_n}, n \geq 1$ and then

$$u_r \subseteq \bigcap_{n=1}^{\infty} u_{r_n}.$$

We will have to prove the converse inclusion. First let us start with the remark that since $(u_m)_r^- \to u_r^-$ and $(u_m)_r^+ \to u_r^+$ for any $\varepsilon_1 > 0$ there exists an $M_1 \geq 1$ such that for $m \geq M_1$ we have

$$[(u_m)_r^-, (u_m)_r^+,] \subseteq [u_r^- - \varepsilon_1, u_r^+ + \varepsilon_1].$$

Since $u_m, m \geq 1$ is a fuzzy number, we have

$$(u_m)_r = \bigcap_{n=1}^{\infty} (u_m)_{r_n}.$$

Also, from $(u_m)_{r_n}^- \to u_{r_n}^-$ and $(u_m)_{r_n}^+ \to u_{r_n}^+$ we have that for any $\varepsilon_2 > 0$ there exists $M_2 \geq 1$ such that for $m \geq M_2$ and any $n \geq 1$,

$$[u_{r_n}^- + \varepsilon_2, u_{r_n}^+ - \varepsilon_2] \subseteq (u_m)_{r_n}.$$

which implies

$$\bigcap_{n=1}^{\infty} [u_{r_n}^- + \varepsilon_2, u_{r_n}^+ - \varepsilon_2] \subseteq \bigcap_{n=1}^{\infty} (u_m)_{r_n}.$$

Now from these relations we obtain that for any $\varepsilon_1, \varepsilon_2 > 0$ we have

$$\bigcap_{n=1}^{\infty} [u_{r_n}^- + \varepsilon_2, u_{r_n}^+ - \varepsilon_2] \subseteq [u_r^- - \varepsilon_1, u_r^+ + \varepsilon_1]$$

which leads to

$$\bigcap_{n=1}^{\infty} u_{r_n} \subseteq u_r.$$

The double inclusion proves (iii) of Negoita-Ralescu characterization theorem.

(iv) Let now $r_n \searrow 0$ be a sequence in $[0, 1]$. We will have to prove that

$$cl \left( \bigcup_{n=1}^{\infty} u_{r_n} \right) = u_0.$$

Since $u_0$ is closed and since $u_{r_n} \subseteq u_0$ the inclusion

$$cl \left( \bigcup_{n=1}^{\infty} u_{r_n} \right) \subseteq u_0$$

is immediate. For the converse we follow a reasoning similar to the previous point. Let $\varepsilon_1, \varepsilon_2 > 0$ be arbitrary. Then we have

$$[u_0^- + \varepsilon_1, u_0^+ - \varepsilon_1] \subseteq (u_m)_0 = cl\left(\bigcup_{n=1}^{\infty} (u_m)_{r_n}\right),$$

with $m \geq M_1$. Also

$$cl\left(\bigcup_{n=1}^{\infty} (u_m)_{r_n}\right) \subseteq cl\left(\bigcup_{n=1}^{\infty} [u_{r_n}^- - \varepsilon_2, u_{r_n}^+ + \varepsilon_2]\right),$$

for $m \geq M_2$. Finally we obtain

$$[u_0^- + \varepsilon_1, u_0^+ - \varepsilon_1] \subseteq cl\left(\bigcup_{n=1}^{\infty} [u_{r_n}^- - \varepsilon_2, u_{r_n}^+ + \varepsilon_2]\right),$$

$\varepsilon_1, \varepsilon_2 > 0$. The required relation

$$cl\left(\bigcup_{n=1}^{\infty} u_{r_n}\right) = u_0$$

follows.

Conditions (i)-(iv) imply that $u$ is a fuzzy number, i.e., the Cauchy sequence $u_n \in \mathbb{R}_{\mathcal{F}}$ converges to $u \in \mathbb{R}_{\mathcal{F}}$ and the proof is complete. ∎

The next theorem shows that the $L^p$-type metric for fuzzy numbers is not generating a complete space of fuzzy numbers, property which makes the $L^p$-type metric difficult to be used.

**Theorem 8.6.** $(\mathbb{R}_{\mathcal{F}}, D_p)$ *is not complete,* $1 \leq p < \infty$.

**Proof.** We will construct a Cauchy sequence that is not convergent within $(\mathbb{R}_{\mathcal{F}}, D_p)$. In this proof we assume $p$ to be an integer, but the result can be extended to real values $1 \leq p < \infty$, using special functions.

Let

$$u_n(x) = \begin{cases} 0 & \text{if } x \notin [0, n] \\ e^{-x} & \text{if } x \in [0, n] \end{cases}, n \geq 1$$

be a sequence of fuzzy numbers. Its level sets are given by

$$(u_n)_r = \begin{cases} [0, n] & \text{if } r < e^{-n} \\ -\ln r & \text{if } r \geq e^{-n} \end{cases}.$$

We have

$$D_p(u_n, u_{n+k}) = \left(\int_0^{e^{-n-k}} k^p dr + \int_{e^{-n-k}}^{e^{-n}} (-\ln r - n)^p dr\right)^{\frac{1}{p}}.$$

Let us denote

$$I_p = \int_{e^{-n-k}}^{e^{-n}} (-\ln r - n)^p dr.$$

We have

$$I_0 = e^{-n} - e^{-n-k}.$$

By integration by parts we obtain

$$I_p = k^p e^{-n-k} + p I_{p-1}.$$

Also,

$$I_{p-1} = k^{p-1} e^{-n-k} + (p-1) I_{p-2}$$

and then

$$I_p = k^p e^{-n-k} + pk^{p-1} e^{-n-k} + p(p-1) I_{p-2}.$$

By induction we have

$$I_p = (k^p + pk^{p-1} + ... + p(p-1)...2 \cdot k) e^{-n-k}$$

$$+ p! \left( e^{-n} - e^{-n-k} \right).$$

Then we have

$$D_p(u_n, u_{n+k}) = \left( k^p e^{-n-k} + (k^p + pk^{p-1} + ... \right.$$

$$\left. + p(p-1)...2 \cdot k) e^{-n-k} + p! \left( e^{-n} - e^{-n-k} \right) \right)^{\frac{1}{p}},$$

that is

$$D_p(u_n, u_{n+k})$$

$$\leq \left( (k^p + (k^p + pk^{p-1} + ... + p(p-1)...2 \cdot k) - p!) e^{-n-k} + p! e^{-n} \right)^{\frac{1}{p}}.$$

If $k < p$ we can estimate this Hausdorff distance by

$$D_p(u_n, u_{n+k})$$

$$\leq \left( (p^p + (p^p + pp^{p-1} + ... + p(p-1)...2 \cdot p) - p!) e^{-k} + p! \right)^{\frac{1}{p}} \cdot e^{-\frac{n}{p}}.$$

Now, if $k \geq p$ then we have

$$D_p(u_n, u_{n+k}) = \left( (k^p + k^{p+1} - p!) e^{-k} + p! \right)^{\frac{1}{p}} \cdot e^{-\frac{n}{p}},$$

and we obtain

$$\lim_{n \to \infty} D_p(u_n, u_{n+k}) = 0.$$

Then it follows that the sequence $u_n$ is a Cauchy sequence. Now we can see that $u_n$ does not converge within $\mathbb{R}_{\mathcal{F}}$. Indeed, it converges point-wise to

$$u(x) = \begin{cases} 0 & \text{if } x < 0 \\ e^{-x} & \text{if } x \geq 0 \end{cases}.$$

Indeed, we have

$$D_p(u_n, u) = \left( \int_0^{e^{-n}} (-\ln r - n)^p dr \right)^{\frac{1}{p}}.$$

We calculate

$$J_p = \int_0^{e^{-n}} (-\ln r - n)^p dr.$$

We have $J_0 = e^{-n}$ and by integration by parts we get

$$J_p = r(-\ln r - n)^p|_0^{e^{-n}} + p \int_0^{e^{-n}} r(-\ln r - n)^{p-1} \frac{1}{r} dr,$$

i.e $J_p = pJ_{p-1}$ which leads to

$$J_p = p! e^{-n}.$$

Finally we obtain

$$\lim_{n \to \infty} D_p(u_n, u) = \lim_{n \to \infty} \left( p! e^{-n} \right)^{\frac{1}{p}} = 0.$$

Since $u \notin \mathbb{R}_{\mathcal{F}}$ we obtain that $\mathbb{R}_{\mathcal{F}}$ is not complete with respect to the metric $D_p$. ∎

## 8.3   Compactness

A metric space is locally compact if every point has a compact neighborhood. This means that we will need to study compact subsets in the metric space of fuzzy numbers when considering the different metrics that we have studied in the previous sections. A subset of a metric space on its turn is compact if and only if it is complete and totally bounded or also, equivalently, if and only if it is sequentially compact. Recall that a subspace is sequentially compact if any sequence has a convergent subsequence. A compact neighborhood in a metric space can be only closed and since it is closed, it contains the closure of an open ball that would be at its turn also compact. In $(\mathbb{R}_{\mathcal{F}}, D_\infty)$ and in $(\mathbb{R}_{\mathcal{F}}, D_p)$ the closed unit ball is the closure of the open unit ball (please note that this is not generally true in metric spaces). So, if we succeed to show that a closed ball is not compact then the space we consider would not be locally compact. So we begin with the following theorem:

**Theorem 8.7.** *The closed unit ball of $(\mathbb{R}_{\mathcal{F}}, D_\infty)$ is not compact.*

**Proof.** The closed unit ball in $\mathbb{R}_{\mathcal{F}}$ is

$$\bar{B}(0,1) = \{u \in \mathbb{R}_{\mathcal{F}} | D_\infty(u, 0) \le 1\},$$

where we consider the singleton 0 coincident with the real number 0. We will construct a sequence without a convergent subsequence. It is known that the rational numbers set $\mathbb{Q}$ is countable. So let us write it as a sequence $q_n \in [0, 1]$ and we define

$$u_n(x) = \begin{cases} 0 & \text{if} \quad x \notin [0, 1] \\ q_n & \text{if} \quad 0 \le x < \frac{1}{2} \\ 2(1 - q_n)x + 2q_n - 1 & \text{if} \quad \frac{1}{2} \le x \le 1 \end{cases}.$$

The level sets of $u_n$ can be written

$$(u_n)_r = \begin{cases} [0, 1] & \text{if} \quad 0 \le r < q_n \\ \left[\frac{r+1-2q_n}{2(1-q_n)}, 1\right] & \text{if} \quad q_n \le r \le 1 \end{cases}.$$

Then the Hausdorff-Pompeiu distance level-wise between two elements $u_n$ and $u_{n+k}$ of the sequence ($q_n < q_{n+k}$ is assumed for simplicity without restricting the generality) is given as

$$d_H((u_n)_r, (u_{n+k})_r) = \begin{cases} 0 & \text{if} \quad 0 \le r < q_n \\ \frac{r+1-2q_n}{2(1-q_n)} & \text{if} \quad q_n \le r < q_{n+k} \\ \frac{r+1-2q_n}{2(1-q_n)} - \frac{r+1-2q_{n+k}}{2(1-q_{n+k})} & \text{if} \quad q_{n+k} \le r \le 1 \end{cases}.$$

We have

$$D_\infty(u_n, u_{n+k}) = \sup_{r \in [0,1]} d_H((u_n)_r, (u_{n+k})_r).$$

i.e.,

$$D_\infty(u_n, u_{n+k}) = \sup_{r \in [q_n, q_{n+k}]} \frac{r + 1 - 2q_n}{2(1 - q_n)}$$

$$= \frac{q_{n+k} + 1 - 2q_n}{2(1 - q_n)} \ge \frac{1}{2}.$$

This last relation means that the distance between any two elements of this sequence is at least $\frac{1}{2}$. Now let us suppose that the sequence $u_n$ has a convergent subsequence. That subsequence would be a Cauchy sequence, which is in contradiction with the fact that any two elements of it are at least $\frac{1}{2}$ apart. This shows that it cannot have a convergent subsequence. So, the unit ball is not sequentially compact and as a consequence it is not compact. ∎

**Remark 8.8.** *It is easy to see that the unit ball is not preferred by compactness properties over balls with a different radius, so in general we can assume that a closed ball is not compact in $D_\infty$. This means that a point in $\mathbb{R}_{\mathcal{F}}$ does not have a compact neighborhood and this also implies that $\mathbb{R}_{\mathcal{F}}$ is not locally compact.*

A similar result can be deduced concerning the $L^p$ type metrics on fuzzy numbers.

**Theorem 8.9.** *The closed unit ball*

$$\bar{B}_p(0,1) = \{u \in \mathbb{R}_{\mathcal{F}} | D_p(u,0) \le 1\}$$

*in* $(\mathbb{R}_{\mathcal{F}}, D_p)$ *is not compact.*

**Proof.** We will prove the result for integer values $1 \le p < \infty$, however the proof can be extended to the case of real values using special functions. As in the proof of the previous Theorem we will construct a sequence of fuzzy numbers in the unit ball of $(\mathbb{R}_{\mathcal{F}}, D_p)$. The construction is similar to the one in Theorem 8.6.

Let

$$u_n(x) = \begin{cases} 0 & \text{if } x \notin [0,n] \\ e^{-(p!)^{\frac{1}{p}}x} & \text{if } x \in [0,n] \end{cases}, n \ge 1$$

be a sequence of fuzzy numbers. Its level sets are given by

$$(u_n)_r = \begin{cases} [0,n] & \text{if } r < e^{-(p!)^{\frac{1}{p}}n} \\ -\dfrac{1}{(p!)^{\frac{1}{p}}} \ln r & \text{if } r \ge e^{-(p!)^{\frac{1}{p}}n} \end{cases}.$$

Let us consider the point-wise limit of the sequence

$$u(x) = \begin{cases} 0 & \text{if } x < 0 \\ e^{-(p!)^{\frac{1}{p}}x} & \text{if } x \ge 0 \end{cases}.$$

It is easy to see that $u_n(x) \le u(x), \forall x \in \mathbb{R}$. The level-wise Hausdorff distance is

$$d_H((u_n)_r, 0) = \begin{cases} n & \text{if } r < e^{-(p!)^{\frac{1}{p}}n} \\ -\dfrac{1}{(p!)^{\frac{1}{p}}} \ln r & \text{if } r \ge e^{-(p!)^{\frac{1}{p}}n} \end{cases}$$

We have

$$D_p(u_n, 0) = \left( \int_0^1 d_H \left((u_n)_r, 0\right)^p dr \right)^{\frac{1}{p}}$$

$$\le \left( \int_0^1 \left( -\frac{1}{(p!)^{\frac{1}{p}}} \ln r \right)^p dr \right)^{\frac{1}{p}}$$

$$= \left( \frac{1}{p!} \int_0^1 (-\ln r)^p dr \right)^{\frac{1}{p}} = \left( \frac{1}{p!} I_p \right)^{\frac{1}{p}}.$$

To compute $I_p$ we integrate by parts and we obtain

$$I_p = r\left(-\ln r\right)^p |_0^1 + p \int_0^1 (-\ln r)^{p-1} dr = p I_{p-1}.$$

This shows that $I_p = p!$ and we obtain $D_p(u_n, 0) \leq 1$ that means $u_n \in \bar{B}_p(0,1) \subseteq \mathbb{R}_{\mathcal{F}}$. Using the same calculation as that in the proof of Theorem 8.6 we obtain that $u_n$ does not converge in $(\mathbb{R}_{\mathcal{F}}, D_p)$. It is easy to check that the sequence is a Cauchy sequence. In a metric space, a Cauchy sequence that has a convergent subsequence is itself convergent, but this is not true about our sequence $u_n$. So, $u_n$ does not have a convergent subsequence. As a conclusion $\bar{B}_p(0,1)$ is not compact. ∎

**Remark 8.10.** *It is easy to see that the conclusion can be generalized and a closed ball is not compact in $D_p$. This means that a point in $\mathbb{R}_{\mathcal{F}}$ does not have a compact neighborhood and this also implies that $\mathbb{R}_{\mathcal{F}}$ is not locally compact with respect to the $D_p$ metric.*

## 8.4 Separability

Separability is another important property mainly for approximation purposes. A metric space is separable if it has a dense countable subset. Based on the same construction ideas as those in the previous sections we will prove that $(\mathbb{R}_{\mathcal{F}}, D_p)$ $1 \leq p < \infty$ are separable spaces while $(\mathbb{R}_{\mathcal{F}}, D_\infty)$ is not separable. Let us suppose that there is a dense countable subset $Y$ of a given metric space $X$. If there would be an open ball that is disjoint from the countable dense subset $Y$ then the center of that open ball cannot be reached by a convergent sequence in $Y$. As a conclusion the center of the open ball we considered would not be in the closure of $Y$ i.e., $Y$ could not be dense in $X$. As a conclusion, to show that a space is not separable it is enough to find a family of uncountably many disjoint open balls in $X$. Then each of them would necessarily contain at least an element of $Y$ so there would be uncountably many elements in the countable dense subset. Of course this is a contradiction. So, to show that a metric space is not separable, it is enough to exhibit uncountably many disjoint open balls.

First we have a positive result about $(\mathbb{R}_{\mathcal{F}}, D_p)$.

**Theorem 8.11.** *The metric space $(\mathbb{R}_{\mathcal{F}}, D_p)$ is separable.*

**Proof.** The proof is following the basic idea of Diamond-Kloeden [44], but it is simplified. We consider a fixed fuzzy number $u \in \mathbb{R}_{\mathcal{F}}$. Since its support is compact it is totally bounded, i.e., we can cover it with finitely many balls of any given radius $\varepsilon > 0$. We can assume that we find a specific minimal covering, provided by rational numbers $r_1 < r_2 < ... < r_n$ as intervals with endpoints at rational numbers, of length $r_{i+1} - r_i < 2\varepsilon$. We have

$$(u)_0 \subseteq \bigcup_{i=1}^{n-1} [r_i, r_{i+1}).$$

Also, in this case we have for any level-set

$$(u)_\alpha \subseteq \bigcup_{i \in I_\alpha} [r_i, r_{i+1}),$$

with $I_\alpha$ minimal. Let us consider now $\alpha_i$ rational numbers such that

$$\alpha_i \leq \sup_{x \in [r_i, r_{i+1}]} u(x) < \alpha_i + \varepsilon.$$

Let us define

$$v(x) = \sum_{i=1}^{n-1} \chi_{[r_i, r_{i+1}]} \alpha_i.$$

It is easy to see that the cardinality of the set of all such functions is the same as that of $\mathbb{Q}$, i.e., it is countable. Also, it is easy to see that

$$(v)_\alpha \subseteq \bigcup_{i \in J_\alpha} [r_i, r_{i+1})$$

where $J_\alpha = \{i | \alpha_i \geq \alpha\}$. It is easy to see that $I_\alpha$ and $J_\alpha$ consist of the same intervals, except possibly they may differ at the end on intervals that have length less than $2\varepsilon$. Let

$$A = \{\alpha \in [0,1] | d_H((u)_\alpha, (v)_\alpha) < \varepsilon\}.$$

We can see that $[0,1] \setminus A \subseteq [r_j, r_{j+1}) \cup [r_k, r_{k+1})$ for some $j, k \in \{1, ..., n\}$. We can conclude

$$D_p(u,v) = \left( \int_0^1 d_H((u)_\alpha, (v)_\alpha)^p d\alpha \right)^{\frac{1}{p}}$$

$$= \left( \int_A d_H((u)_\alpha, (v)_\alpha)^p d\alpha \right)^{\frac{1}{p}}$$

$$+ \left( \int_{[0,1] \setminus A} d_H((u)_\alpha, (v)_\alpha)^p d\alpha \right)^{\frac{1}{p}}$$

$$< \varepsilon + 4\varepsilon \cdot \operatorname{diam}(u)_0.$$

As a conclusion we find an element $v$ of a countable set in any neighborhood of a fuzzy number $u \in \mathbb{R}_\mathcal{F}$ with respect to the distance $D_p$ and this completes the proof. ∎

Now let us prove that $(\mathbb{R}_\mathcal{F}, D_\infty)$ is not separable.

**Theorem 8.12.** *The closed unit ball*

$$\bar{B}(0,1) = \{u \in \mathbb{R}_\mathcal{F} | D_\infty(u, 0) \leq 1\}$$

*in $(\mathbb{R}_\mathcal{F}, D_\infty)$ is not separable.*

**Proof.** The construction is similar to that in Theorem 8.7. For a given $t \in [0, 1]$ we define

$$
u_t(x) = \begin{cases} 0 & \text{if } x \notin [0, 1] \\ t & \text{if } 0 \leq x < \frac{1}{2} \\ 2(1 - t)x + 2t - 1 & \text{if } \frac{1}{2} \leq x \leq 1 \end{cases}.
$$

The level sets of $u_t$ can be written

$$
(u_t)_r = \begin{cases} [0, 1] & \text{if } 0 \leq r < t \\ \left[ \frac{r+1-2t}{2(1-t)}, 1 \right] & \text{if } t \leq r \leq 1 \end{cases}.
$$

Then the Hausdorff-Pompeiu distance level-wise between two elements $u_t$ and $u_s$ of the sequence, $t < s$ is given as

$$
d_H((u_t)_r, (u_s)_r) = \begin{cases} 0 & \text{if } 0 \leq r < s \\ \frac{r+1-2t}{2(1-t)} & \text{if } t \leq r < s \\ \frac{r+1-2t}{2(1-t)} - \frac{r+1-2s}{2(1-s)} & \text{if } s \leq r \leq 1 \end{cases}.
$$

We have

$$
D_\infty(u_t, u_s) = \sup_{r \in [t,s]} \frac{r + 1 - 2t}{2(1 - t)} \geq \frac{1}{2} > \frac{1}{3}.
$$

So the open balls $B\left(u_t, \frac{1}{3}\right)$ are disjoint and uncountably many. So a countable dense subset if there would exist, would need to have an element in each such ball, which is impossible. So, the closed unit ball in $(\mathbb{R}_\mathcal{F}, D_\infty)$ is not separable. ∎

It is known that a metric space is separable if and only if it is Lindelöf (i.e., any open covering has a countable sub-cover). We can conclude that the space $(\mathbb{R}_\mathcal{F}, D_\infty)$ is not a Lindelöf space, while $(\mathbb{R}_\mathcal{F}, D_p)$ is Lindelöf.

As a consequence we can mention two of the problems which should be avoided when we want to use the metrics over the fuzzy reals. First, if we have a Cauchy sequence in $(\mathbb{R}_\mathcal{F}, D_p)$, for $1 \leq p < \infty$ we cannot be sure about its convergence. In $(\mathbb{R}_\mathcal{F}, D_\infty)$ we cannot use a density argument (we cannot reduce a problem to a sub-problem of a countable dimension). Finally in both $(\mathbb{R}_\mathcal{F}, D_\infty)$ and $(\mathbb{R}_\mathcal{F}, D_p)$ we face serious difficulties when we want to extract a convergent subsequence of a bounded sequence.

The fuzzy analysis literature is using mostly the Hausdorff distance between the fuzzy numbers, because in this case the structure of the metric space $(\mathbb{R}_\mathcal{F}, D_\infty)$ being complete, is near to the structure of a Banach space. Surely we do not have a linear space structure, so it cannot be a Banach space, but several properties which hold in Banach spaces also hold in $(\mathbb{R}_\mathcal{F}, D_\infty)$.

## 8.5   Norm of a Fuzzy Number

Let us denote $\|u\|_{\mathcal{F}} = D_\infty(u, 0)$, $\forall u \in \mathbb{R}_{\mathcal{F}}$ the norm of a fuzzy number. The following theorem shows us that this has properties similar to the properties of a norm in the usual crisp sense, without being a norm. It is not a norm because $\mathbb{R}_{\mathcal{F}}$ is not a linear space.

**Proposition 8.13.** *(Anastassiou-Gal [5], Gal [69])* $\|\cdot\|_{\mathcal{F}}$ *has the following properties*

(i) $\|u\|_{\mathcal{F}} = 0$ *if and only if* $u = 0$;

(ii) $\|\lambda \cdot u\|_{\mathcal{F}} = |\lambda| \cdot \|u\|_{\mathcal{F}}$, $\forall \lambda \in \mathbb{R}$ *and* $u \in \mathbb{R}_{\mathcal{F}}$;

(iii) $\|u + v\|_{\mathcal{F}} \leq \|u\|_{\mathcal{F}} + \|v\|_{\mathcal{F}}$, $\forall u, v \in \mathbb{R}_{\mathcal{F}}$;

(iv) $|\|u\|_{\mathcal{F}} - \|v\|_{\mathcal{F}}| \leq D_\infty(u, v)$, $\forall u, v \in \mathbb{R}_{\mathcal{F}}$;

(v) *For any a and b having the same sign and any* $u \in \mathbb{R}_{\mathcal{F}}$ *we have*

$$D_\infty(a \cdot u, b \cdot u) = |b - a| \cdot \|u\|_{\mathcal{F}};$$

(vi) $D_\infty(u, v) = \|u \ominus_{gH} v\|_{\mathcal{F}}$, $\forall u, v \in \mathbb{R}_{\mathcal{F}}$.

**Proof.** The proofs of (i)-(iv) and (vi) are left to the reader as an exercise. We will prove (v). Let $u \in \mathbb{R}_{\mathcal{F}}$, and $a > b > 0$. Then we have

$$D_\infty(a \cdot u, b \cdot u) = D_\infty([b + (a - b)] \cdot u, b \cdot u) = D_\infty(b \cdot u + (a - b) \cdot u, b \cdot u).$$

Since $D_\infty$ is invariant to translations we get

$$D_\infty(a \cdot u, b \cdot u) = D_\infty((a - b) \cdot u, 0) = |b - a| \cdot \|u\|_{\mathcal{F}}.$$ ∎

**Remark 8.14.** *The concepts of limit, convergence of a sequence in* $\mathbb{R}_{\mathcal{F}}$ *and continuity of a function* $f : [a, b] \to \mathbb{R}_{\mathcal{F}}$ *will all be considered in the metric space* $(\mathbb{R}_{\mathcal{F}}, D_\infty)$. *For example continuity of a fuzzy number valued function at a point* $x_0$ *means that* $\forall \varepsilon > 0, \exists \delta > 0$ *such that* $D_\infty(f(x), f(x_0)) < \varepsilon$ *for* $|x - x_0| < \delta$. *From the definition of the Hausdorff distance this implies*

$$\sup_{r \in [0,1]} \max\{|f_r^-(x) - f_r^-(x_0)|, |f_r^+(x) - f_r^+(x_0)|\} < \varepsilon$$

*which results in* $|f_r^\pm(x) - f_r^\pm(x_0)| < \varepsilon$, $\forall r \in [0, 1]$, *i.e., equicontinuity of the family* $\{f_r^\pm | r \in [0, 1]\}$. *In particular it also means that these functions are continuous.*

The algebraic-analytic properties of fuzzy number's space makes them very interesting. Recently in Gal [69] it was shown that the general properties shown in Theorem 5.8, and 8.3 allow a generalization in a natural way yielding the definition and study of the following abstract space.

**Definition 8.15.** *We say that* $(X, +, \cdot, d)$ *is a fuzzy-number type space (shortly FN-type space), if the following properties are satisfied :*

*(i)* $(X, d)$ *is a metric space (complete or not) and d has the properties in Theorem 8.3.*

*(ii) The operations* $+, \cdot$ *on* $X$ *have the properties in Theorem 5.8, (i),(iv), (v),(vi);*

*(iii) There exists a neutral element* $0 \in X$, *i.e.* $u + 0 = 0 + u = u$, *for any* $u \in X$ *and a subspace* $Y \subset X$ *(with respect to* $+$ *and* $\cdot$*), non-dense in* $X$, *such that with respect to* $0$, *none of* $u \in X \setminus Y$ *has an opposite element in* $X$.

Obviously $(\mathbb{R}_{\mathcal{F}}, D_\infty)$, $(\mathbb{R}_{\mathcal{F}}, D_p)$ are both FN-type spaces. The uniform distance is defined as in the usual case for continuous functions $f, g : [a, b] \to \mathbb{R}_{\mathcal{F}}$

$$D(f, g) = \sup\{D_\infty(f(x), g(x)|x \in [a, b]\}.$$

Then $(C[a, b], D)$ is an FN-type space. Indeed, because $(\mathbb{R}_{\mathcal{F}}, D)$ is a complete metric space, it is easy to prove that $(C([a, b]; \mathbb{R}_{\mathcal{F}}), D)$ is a complete metric space. Also, if we define

$$(f + g)(x) = f(x) + g(x),$$

$$(\lambda \cdot f)(x) = \lambda \cdot f(x)$$

(for simplicity, the addition and scalar multiplication in $C([a, b]; \mathbb{R}_{\mathcal{F}})$ are denoted as in $\mathbb{R}_{\mathcal{F}}$), also $0 : [a, b] \to \mathbb{R}_{\mathcal{F}}, 0(t) = 0$, for all $t \in [a, b]$,

$$\|f\|_{\mathcal{F}} = \sup\{D(0, f(x)); x \in [a, b]\} .$$

Other examples and several new properties were shown in Gal [69].


## 8.6   Embedding Theorem for Fuzzy Numbers

In the followings we will discuss an embedding result concerning fuzzy numbers.

Let $\overline{C}[0, 1] = \{F : [0, 1] \to \mathbb{R}; F$ bounded on $[0, 1]$, left continuous for $x \in (0, 1]$, right continuous at $0\}$. Together with the norm

$$\|F\| = \sup\{|F(x)|; x \in [0, 1]\},$$

$\overline{C}[0, 1]$ is a Banach space. We can embed $\mathbb{R}_{\mathcal{F}}$ into the Banach space $\overline{C}[0, 1]^2$ according to the following theorem:

**Theorem 8.16.** *(Ma [104]) Let*

$$j : \mathbb{R}_{\mathcal{F}} \to \overline{C}[0, 1] \times \overline{C}[0, 1]$$

*given by*

$$j(u) = (u^-, u^+),$$

*where* $u^-, u^+ : [0, 1] \to \mathbb{R}$, $u^-(r) = u_r^-$, $u^+(r) = u_r^+$. *Then* $j(\mathbb{R}_{\mathcal{F}})$ *is a closed convex cone having its vertex at 0 in* $\overline{C}[0, 1] \times \overline{C}[0, 1]$. *Here* $\overline{C}[0, 1] \times \overline{C}[0, 1]$ *is a Banach space w.r.t. the norm*

$$\|(f, g)\| = \max\{\|f\|, \|g\|\}.$$

*Moreover, $j$ satisfies:*

*(i)*

$$j(a \cdot u + b \cdot v) = aj(u) + bj(v),$$

$\forall u, v \in \mathbb{R}_{\mathcal{F}}$, $a, b \geq 0$;

*(ii)*

$$D_\infty(u, v) = \|j(u) - j(v)\|.$$

**Proof.** The representation of a fuzzy number through the functions $u^-$, $u^+$ in Theorem 4.10 ensures that $j$ is well defined and injective. We can deduce based on the same theorem, that

$$j(\mathbb{R}_{\mathcal{F}}) = \{(u^-, u^+) \in \overline{C}[0, 1]^2 | u^- \text{-nondecreasing},$$

$$u^+ \text{-nonincresing}, u_1^- \leq u_1^+\}.$$

To check that it is closed it is enough to observe that taking a sequence of elements in $j(\mathbb{R}_{\mathcal{F}})$ has monotone components that converge to a monotone limit in the uniform norm. Also, convergence keeps the relation $u_1^- \leq u_1^+$ intact and so, the space $j(\mathbb{R}_{\mathcal{F}})$ is closed. Considering now

$$(u^-, u^+), (v^-, v^+) \in j(\mathbb{R}_{\mathcal{F}})$$

and $a, b > 0$ we have

$$j(a \cdot u + b \cdot v) = a(u^-, u^+) + b(v^-, v^+) = aj(u) + bj(v),$$

which proves $(i)$ and also the fact that $j(\mathbb{R}_{\mathcal{F}})$ is a convex cone. To prove $(ii)$ we observe

$$\|j(u) - j(v)\| = \max\{\sup_{r \in [0,1]} |u_r^- - v_r^-|, \sup_{r \in [0,1]} |u_r^+ - v_r^+|\} = D_\infty(u, v),$$

and the proof is complete.                                                ∎

## 8.7   Fuzzy Numbers with Continuous Endpoints of the Level Sets

Let us now consider the set of fuzzy numbers such that the left and right endpoint of the level set functions are continuous.

$$\mathbb{R}_{\mathcal{F}}^c = \{u \in \mathbb{R}_{\mathcal{F}} | u_r = [u_r^-, u_r^+], u_r^\pm \in C[0, 1]\}.$$

Let us observe that this space does not coincide with the space of fuzzy numbers with continuous membership degrees. Indeed, for example singletons, and crisp intervals regarded as fuzzy sets through their characteristic functions

$$u(x) = \begin{cases} 1 & \text{if} \quad x \in [a, b] \\ 0 & \text{otherwise} \end{cases}$$

are discontinuous while the level-set functions associated to them will be continuous. The level set functions in this case being constant $u_r^- = a, u_r^+ = b, r \in [0, 1]$ they are continuous, while the characteristic function is discontinuous. Also, there are fuzzy numbers with continuous membership function such that the level set functions $u_r^\pm$ are discontinuous. It is enough to consider $u : [0, 4] \to [0, 1]$,

$$u(x) = \begin{cases} \frac{x}{2} & \text{if} \quad x \in [0, 1) \\ \frac{1}{2} & \text{if} \quad x \in [1, 2] \\ \frac{x-1}{2} & \text{if} \quad x \in [2, 3] \\ 4 - x & \text{if} \quad x \in [3, 4] \end{cases},$$

and we have $u_{\frac{1}{2}}^- = 1$ while $\lim_{r \searrow \frac{1}{2}} (u_r^-) = 2$. As we can see restriction to continuous endpoints of the level set is not too restrictive.

Another research direction (see Rojas-Medar and Roman-Flores [126] Barros-Bassanezi-Tonelli [12]), which is leading to a larger space within fuzzy numbers space is to consider those fuzzy numbers which do not have a proper local minimum of their membership function, i.e., there is no $0 < x_0 < 1$ such that $u(x_0) \leq u(x)$ in a neighborhood $x \in [x_0 - \varepsilon, x_0 + \varepsilon]$ and their core consists of a single point. The subset of fuzzy numbers obtained by these restrictions will be separable and complete.

In most of the situations we try to address the more general space of fuzzy numbers without any restriction.

**Proposition 8.17.** $\mathbb{R}_{\mathcal{F}}^c$ *with* $D_\infty$ *is a separable, complete metric space.*

**Proof.** First let us observe that being a closed subspace of a complete metric space, $\mathbb{R}_{\mathcal{F}}^c$ is complete. Let us consider $C[0, 1]$ the Banach space of continuous real-valued functions on $[0, 1]$ and the embedding

$$j : \mathbb{R}_{\mathcal{F}}^c \to C[0, 1] \times C[0, 1]$$

given by

$$j(u) = (u^-, u^+),$$

where $u^-, u^+ : [0, 1] \to \mathbb{R}$, $u^-(r) = u_r^-$, $u^+(r) = u_r^+$. Then $j(\mathbb{R}_{\mathcal{F}})$ is a closed convex cone having its vertex at 0 in $C[0, 1] \times C[0, 1]$ with the norm

$$\|(f, g)\| = \max\{\|f\|, \|g\|\}.$$

The embedding $j$ is the same as the one defined in an earlier section but its image is now within the space of continuous functions. Being isometrically embedded in a subspace of $C[0,1]$ each component of $j(\mathbb{R}_\mathcal{F}^c)$ is separable according to the Banach-Mazur theorem. Then as a consequence itself $j(\mathbb{R}_\mathcal{F}^c)$ is separable and so is $\mathbb{R}_\mathcal{F}^c$ itself.    ∎

**Remark 8.18.** *It is well known that the closed unit ball in $C[0,1]$ is not compact, so clearly the closed unit ball in $\mathbb{R}_\mathcal{F}^c$ will not be compact. Boundedness and continuity of elements of a subset of a function space is insufficient for compactness. It is replaced by equicontinuity in e.g. Arzela-Ascoli type results.*

## 8.8   Integration of Fuzzy-Number-Valued Functions

In is easy to observe that the definition of the integral of a fuzzy-number-valued function does not raise serious problems. In the followings Aumann, Riemann and Henstock type integrals will be introduced.

First we introduce an Aumann-type integral, used in several works.

**Definition 8.19.** *(Diamond-Kloeden [44]) A mapping $f : [a,b] \to \mathbb{R}_\mathcal{F}$ is said to be strongly measurable if the level set mapping $[f(x)]_\alpha$ are measurable for all $\alpha \in [0,1]$. Here measurable means Borel measurable.*

*A fuzzy-valued mapping $f : [a,b] \to \mathbb{R}_\mathcal{F}$ is called integrably bounded if there exists an integrable function $h : [a,b] \to \mathbb{R}$, such that*

$$\|f(t)\|_\mathcal{F} \leq h(t), \ \forall t \in [a,b]$$

*A strongly measurable and integrably bounded fuzzy-valued function is called integrable. The **fuzzy Aumann integral** of $f : [a,b] \to \mathbb{R}_\mathcal{F}$ is defined levelwise by the equation*

$$\left[ (FA) \int_a^b f(x)\,dx \right]^r = \int_a^b [f(x)]^r\,dx, \ r \in [0,1].$$

The following Riemann type integral presents an alternative to Aumann-type definition.

**Definition 8.20.** *(Gal [66]) A function $f : [a,b] \to \mathbb{R}_\mathcal{F}$, $[a,b] \subset \mathbb{R}$ is called Riemann integrable on $[a,b]$, if there exists $I \in \mathbb{R}_\mathcal{F}$, with the property: $\forall \varepsilon > 0$, $\exists \delta > 0$, such that for any division of $[a,b]$, $d : a = x_0 < ... < x_n = b$ of norm $\nu(d) < \delta$, and for any points $\xi_i \in [x_i, x_{i+1}]$, $i = 0, ..., n-1$, we have*

$$D_\infty \left( \sum_{i=0}^{n-1} f(\xi_i)(x_{i+1} - x_i), I \right) < \varepsilon.$$

*Then we denote $I = (FR) \int_a^b f(x)\,dx$ and it is called **fuzzy Riemann integral**.*

In Wu-Gong [151] the Henstock integral of a fuzzy-valued function is introduced. Surely as a particular case the Riemann-type integral of a fuzzy-valued
function can be re-obtained.

**Definition 8.21.** *(Wu-Gong [151], Bede-Gal [19])Let $f : [a, b] \to \mathbb{R}_{\mathcal{F}}$ a
fuzzy-number-valued function and*

$$\Delta_n : a = x_0 < x_1 < ... < x_{n-1} < x_n = b$$

*a partition of the interval $[a, b]$, $\xi_i \in [x_i, x_{i+1}]$, $i = 0, 1, ..., n - 1$, a sequence
of points of the partition $\Delta_n$ and $\delta(x) > 0$ a real-valued function over $[a, b]$.
The division $P = (\Delta_n, \xi)$ is said to be $\delta$-fine if*

$$[x_i, x_{i+1}] \subseteq (\xi_i - \delta(\xi_i), \xi_i + \delta(\xi_i)).$$

*The function $f$ is said to be Henstock (or (FH)-) integrable having the integral
$I \in \mathbb{R}_{\mathcal{F}}$ if for any $\varepsilon > 0$ there exists a real-valued function $\delta$, such that for
any $\delta$-fine division $P$ we have*

$$D_\infty \left( \sum_{i=0}^{n-1} f(\xi_i) \cdot h_i, I \right) < \varepsilon,$$

*where $h_i = x_{i+1} - x_i$. Then $I$ is called the fuzzy Henstock integral of $f$ and
it is denoted by $(FH) \int_a^b f(t) dt$.*

**Proposition 8.22.** *A continuous fuzzy-number-valued function is fuzzy Aumann integrable, fuzzy Riemann integrable and fuzzy Henstock integrable too,
and moreover*

$$(FA) \int_a^b f(x) dx = (FH) \int_a^b f(x) dx = (FR) \int_a^b f(x) dx.$$

**Proof.** It is immediate to observe that

$$\left[ (FA) \int_a^b f(x) dx \right]^r = \left[ \int_a^b f_-^r(x) dx, \int_a^b f_+^r(x) dx \right], \forall r \in [0, 1].$$

If $f$ is Riemann integrable then it is also Henstock integrable. Indeed, if the
function $\delta$ is constant in the Henstock definition, it will generate the Riemann
case. The Riemann sum can be written level-wise

$$\left[ \sum_{i=0}^{n-1} f(\xi_i)(x_{i+1} - x_i) \right]_r$$

$$= \left[ \sum_{i=0}^{n-1} f_r^-(\xi_i)(x_{i+1} - x_i), \sum_{i=0}^{n-1} f_r^+(\xi_i)(x_{i+1} - x_i) \right].$$

Equicontinuity implies integrability of the functions $f_r^-$ and $f_r^+$ uniformly w.r.t. $r \in [0,1]$. Then we obtain

$$\left[ (FR) \int_a^b f(x) \, dx \right]^r = \left[ \int_a^b f_-^r(x) \, dx, \int_a^b f_+^r(x) \, dx \right],$$

and combining the above results we obtain the equality of the three types of integrals. ∎

The common value of these integrals for a continuous function $f : [a, b] \to \mathbb{R}_{\mathcal{F}}$ will be denoted by $\int_a^b f(x) \, dx$. For general $f : [a, b] \to \mathbb{R}_{\mathcal{F}}$ the above assertion does not hold.

The properties of the integrals for fuzzy functions are similar to the properties of their classical counterparts.

**Proposition 8.23.** *(Gal [66]) The fuzzy integral has the following properties.*

*(i) If $f, g : [a, b] \to \mathbb{R}_{\mathcal{F}}$ are integrable and $\alpha, \beta \in \mathbb{R}$ we have*

$$\int_a^b (\alpha f(x) + \beta g(x)) dx = \alpha \int_a^b f(x) dx + \beta \int_a^b g(x) dx.$$

*(ii) If $f : [a, b] \to \mathbb{R}_{\mathcal{F}}$ is integrable and $c \in [a, b]$, then*

$$\int_a^c f(x) dx + \int_c^b f(x) dx = \int_a^b f(x) dx.$$

*(iii) If $c \in \mathbb{R}_{\mathcal{F}}$ and $f : [a, b] \to \mathbb{R}$ has constant sign on $[a, b]$, then*

$$\int_a^b c \cdot f(x) dx = c \int_a^b f(x) dx.$$

*For general $f$ the property does not hold.*

**Proof.** The proof is left as an exercise. ∎

**Example 8.24.** *(Bede [14]) Let us consider the Green function on $[0, 1]$ interval*

$$G(t, s) = \begin{cases} -s(1 - t), & s \le t \\ -t(1 - s), & s > t \end{cases}.$$

*Then*

$$y(t) = \int_0^t G(t, s)(0, 1, 2) ds + \int_t^1 G(t, s)(0, 1, 2) ds$$

$$= \int_0^t (-2s(1 - t), -s(1 - t), 0) ds + \int_t^1 (-2t(1 - s), -t(1 - s), 0) ds.$$

*We can integrate the terms level-wise and we obtain*

$$y(t) = \left( t^2 - t, \frac{1}{2}(t^2 - t), 0 \right).$$

## 8.9   Differentiability of Fuzzy-Number-Valued Functions

The Hukuhara derivative of a fuzzy-number-valued function was introduced in Puri-Ralescu [123] and it has its starting point in the Hukuhara derivative of multivalued functions. The approach based on the Hukuhara derivative has the disadvantage that a differentiable function has increasing length of its support interval (Diamond [45]). This is not always a realistic assumption. Strongly generalized differentiability of fuzzy-number-valued functions is introduced and studied in Bede-Gal [20]. In this case a differentiable function may have decreasing length of its support. Recently this line of research was continued by introducing gH-derivative and the g-derivative (Bede-Stefanini [27]) of a fuzzy-valued function.

### 8.9.1   Hukuhara Differentiability

**Definition 8.25.** *(Puri-Ralescu [123], Hukuhara [79]) A function $f : (a, b) \to \mathbb{R}_\mathcal{F}$ is called Hukuhara differentiable if for $h > 0$ sufficiently small the the H-differences $f(x + h) \ominus f(x)$ and $f(x) \ominus f(x - h)$ exist and if there exist an element $f'(x) \in \mathbb{R}_\mathcal{F}$ such that*

$$\lim_{h \searrow 0} \frac{f(x + h) \ominus f(x)}{h} = \lim_{h \searrow 0} \frac{f(x) \ominus f(x - h)}{h} = f'(x).$$

*The fuzzy number $f'(x)$ is called the **Hukuhara derivative** of $f$ at $x$.*

**Definition 8.26.** *(Seikkala [131]) The Seikkala derivative of a fuzzy-number-valued function $f : (a, b) \to \mathbb{R}_\mathcal{F}$ is defined by*

$$f'(x)_r = [(f_r^-(x))', (f_r^+(x))'],$$

$0 \leq r \leq 1$, *provided that it defines a fuzzy number $f'(x) \in \mathbb{R}_\mathcal{F}$.*

**Remark 8.27.** *Suppose that the functions $f_r^-(x)$ and $f_r^+(x)$ are continuously differentiable with respect to $x$, uniformly with respect to $r \in [0, 1]$. Then $f$ is Hukuhara differentiable if and only if it is Seikkala differentiable and the two derivatives coincide. Indeed, if $f$ is Hukuhara differentiable we can write*

$$\lim_{h \searrow 0} \frac{f(x + h) \ominus f(x)}{h}$$

$$= \left[ \lim_{h \searrow 0} \frac{f_r^-(x + h) - f_r^-(x)}{h}, \lim_{h \searrow 0} \frac{f_r^+(x + h) - f_r^+(x)}{h} \right].$$

*relation that shows that $f$ is Seikkala differentiable.*
   *Reciprocally if $f$ is Seikkala differentiable then we have its length*

$$len(f(x)_r) = f_r^+(x) - f_r^-(x) \geq 0,$$

*which implies the existence of the Hukuhara difference $f(x + h) \ominus f(x)$ and considering the limit with $h \searrow 0$ we get Hukuhara differentiability. The equality of the two derivatives is immediate.*

**Proposition 8.28.** *A Seikkala differentiable function has non-decreasing length of the closure of the support.*

**Proof.** Indeed, by supposing the contrary, let $f : (a, b) \to \mathbb{R}_{\mathcal{F}}$ be such that $len(f(x))$ is decreasing for some neighborhood of $x \in (a, b)$. Then we have

$$[f'(x)]_0 = [(f_0^-(x))', (f_0^+(x))'].$$

Since $len(f(x)) = f_0^+(x) - f_0^-(x)$ is decreasing we obtain a contradiction. Moreover, if for some $x \in (a, b)$, $f'(x) \in \mathbb{R}_{\mathcal{F}} \setminus \mathbb{R}$ then $len(f(x))$ is strictly increasing. ∎

**Example 8.29.** *(i) If*

$$f(t) = (x(t), y(t), z(t))$$

*is triangular number-valued function, then if $u$ is Hukuhara differentiable and $x, y, z$ are real-valued differentiable functions then*

$$f'(t) = (x'(t), y'(t), z'(t))$$

*is a triangular fuzzy number.*
*(ii) Let $f(t) = (-e^t, 0, e^t)$. Then $f'(t) = (-e^t, 0, e^t)$. And in this case, it is easy to see that $f'(t) = f(t)$ but also, $f'(t) = -f(t)$.*
*(iii) Let $f(t) = (1, 2, 3)e^{-t}$. Suppose that $f(t)$ is differentiable. Then we would have $f'(t) = (-e^{-t}, -2e^{-t}, -3e^{-t})$ which is not a fuzzy number.*

The following example shows a disadvantage of the Hukuhara derivative.

**Example 8.30.** *Let $c \in \mathbb{R}_{\mathcal{F}}$ and $g : (a, b) \to \mathbb{R}_+$ be differentiable on $x_0 \in (a, b)$ and let us define $f : (a, b) \to \mathbb{R}_{\mathcal{F}}$ by*

$$f(x) = c \cdot g(x), x \in (a, b).$$

*Let us suppose that $g'(x_0) > 0$. Then for $h > 0$ sufficiently small, since $g$ is increasing on $[x_0, x_0 + h]$ we obtain*

$$g(x_0 + h) - g(x_0) = \omega(x_0, h) > 0.$$

*Multiplying by $c$, it follows*

$$c \cdot g(x_0 + h) = c \cdot g(x_0) + c \cdot \omega(x_0, h),$$

*i.e. there exists the H-difference $f(x_0 + h) \ominus f(x_0)$. Similarly, reasoning as above we get that there exists the H-difference $f(x_0) \ominus f(x_0 - h)$ too. Also, simple reasoning shows in this case that $f'(x_0) = c \cdot g'(x_0)$. Now, if we suppose*

$g'(x_0) < 0$, *we easily see that we cannot use the above kind of reasoning to prove that the H-differences* $f(x_0 + h) \ominus f(x_0)$, $f(x_0) \ominus f(x_0 - h)$ *and the derivative* $f'(x_0)$ *exist. Consequently, we cannot say that* $f'(x_0)$ *exists. Moreover, we can easily prove that if* $g'(x_0) < 0$ *then the Hukuhara differences* $f(x_0 + h) \ominus f(x_0)$, $f(x_0) \ominus f(x_0 - h)$ *cannot exist, so the function is not Hukuhara differentiable.*

## 8.9.2  *Generalized Differentiabilities*

The definition of strongly generalized differentiability was introduced in Bede-Gal [18], Bede-Gal [20].

**Definition 8.31.** *(Bede-Gal [20]) Let* $f : (a, b) \to \mathbb{R}_{\mathcal{F}}$ *and* $x_0 \in (a, b)$. *We say that* $f$ *is **strongly generalized differentiable** at* $x_0$, *if there exists an element* $f'(x_0) \in \mathbb{R}_{\mathcal{F}}$, *such that*
    *(i) for all* $h > 0$ *sufficiently small,* $\exists f(x_0 + h) \ominus f(x_0)$, $f(x_0) \ominus f(x_0 - h)$ *and the limits (in the metric* $D$)

$$\lim_{h \searrow 0} \frac{f(x_0 + h) \ominus f(x_0)}{h} = \lim_{h \searrow 0} \frac{f(x_0) \ominus f(x_0 - h)}{h} = f'(x_0),$$

*or*
    *(ii) for all* $h > 0$ *sufficiently small,* $\exists f(x_0) \ominus f(x_0 + h)$, $f(x_0 - h) \ominus f(x_0)$ *and the limits*

$$\lim_{h \searrow 0} \frac{f(x_0) \ominus f(x_0 + h)}{(-h)} = \lim_{h \searrow 0} \frac{f(x_0 - h) \ominus f(x_0)}{(-h)} = f'(x_0),$$

*or*
    *(iii) for all* $h > 0$ *sufficiently small,* $\exists f(x_0 + h) \ominus f(x_0)$, $f(x_0 - h) \ominus f(x_0)$ *and the limits*

$$\lim_{h \searrow 0} \frac{f(x_0 + h) \ominus f(x_0)}{h} = \lim_{h \searrow 0} \frac{f(x_0 - h) \ominus f(x_0)}{(-h)} = f'(x_0),$$

*or*
    *(iv) for all* $h > 0$ *sufficiently small,* $\exists f(x_0) \ominus f(x_0 + h)$, $f(x_0) \ominus f(x_0 - h)$ *and the limits*

$$\lim_{h \searrow 0} \frac{f(x_0) \ominus f(x_0 + h)}{(-h)} = \lim_{h \searrow 0} \frac{f(x_0) \ominus f(x_0 - h)}{h} = f'(x_0).$$

*(*$h$ *and* $(-h)$ *at denominators mean* $\frac{1}{h}\cdot$ *and* $-\frac{1}{h}\cdot$, *respectively)*

Case (i) of the previous definition corresponds to the Hukuhara derivative discussed in the previous section.

**Remark 8.32.** *1) This definition is not contradictory, i.e. if for $f$ and $x_0$, at least two from the possibilities (i)-(iv) simultaneously hold, then we do not obtain a contradiction. Indeed, let us suppose, for example, that (i) and (iii) hold. Then by*

$$f(x_0 + h) = f(x_0) + A,$$

$$f(x_0) = f(x_0 - h) + B,$$

$$f(x_0 - h) = f(x_0) + C,$$

*with $A, B, C \in \mathbb{R}_\mathcal{F}$, we get $f(x_0) = f(x_0) + B + C$ but this means $B + C = 0$, which implies $B = C = 0$, i.e., $f'(x_0) = 0$, or $B, C \in \mathbb{R}$, $B = -C$, case when $f'(x_0) \in \mathbb{R}$. But in all these cases it is easy to see that all the limits in the previous definition are equal. Similar conclusion follows for any other combination from (i)-(iv).*

*2) Let $f : (a, b) \to \mathbb{R}_\mathcal{F}$ be strongly generalized differentiable on each point $x \in (a, b)$ in the sense of Definition 8.31, (iii) or (iv). Then $f'(x) \in \mathbb{R}$ for all $x \in (a, b)$.*

**Example 8.33.** *If $g : (a, b) \to \mathbb{R}$ is differentiable on $(a, b)$ such that $g'$ has at most a finite number of roots in $(a, b)$ and $c \in \mathbb{R}_\mathcal{F}$, then $f(x) = c \cdot g(x)$ is strongly generalized differentiable on $(a, b)$ and $f'(x) = c \cdot g'(x)$, $\forall x \in (a, b)$. Indeed, if $g'(x) > 0$, as we have seen above the function is Hukuhara differentiable and so it is strongly generalized differentiable too. If $g'(x) < 0$ then the differences $f(x_0) \ominus f(x_0 + h)$, $f(x_0 - h) \ominus f(x_0)$ exist, and the (ii) case of the above definition 8.31 is satisfied. As $g'(x) = 0$ may involve a sign change, it results in a case (iii) or (iv) of the above definition.*

The concept in Definition 8.31 can be further generalized as follows.

**Definition 8.34.** *(Bede-Gal [20]) Let $f : (a, b) \to \mathbb{R}_\mathcal{F}$ and $x_0 \in (a, b)$. For a sequence $h_n \searrow 0$ and $n_0 \in \mathbb{N}$, let us denote*

$$A_{n_0}^{(1)} = \left\{ n \geq n_0; \exists E_n^{(1)} := f(x_0 + h_n) \ominus f(x_0) \right\},$$

$$A_{n_0}^{(2)} = \left\{ n \geq n_0; \exists E_n^{(2)} := f(x_0) \ominus f(x_0 + h_n) \right\},$$

$$A_{n_0}^{(3)} = \left\{ n \geq n_0; \exists E_n^{(3)} := f(x_0) \ominus f(x_0 - h_n) \right\},$$

$$A_{n_0}^{(4)} = \left\{ n \geq n_0; \exists E_n^{(4)} := f(x_0 - h_n) \ominus f(x_0) \right\}.$$

*We say that $f$ is **weakly generalized differentiable** on $x_0$, if for any sequence $h_n \searrow 0$, there exists $n_0 \in \mathbb{N}$, such that*

$$A_{n_0}^{(1)} \cup A_{n_0}^{(2)} \cup A_{n_0}^{(3)} \cup A_{n_0}^{(4)} = \{ n \in \mathbb{N}; n \geq n_0 \}$$

*and moreover, there exists an element in $\mathbb{R}_\mathcal{F}$ denoted by $f'(x_0)$, such that if for some $j \in \{1, 2, 3, 4\}$ we have $\operatorname{card}(A_{n_0}^{(j)}) = +\infty$, then*

$$\lim_{\substack{h_n \searrow 0 \\ n \to \infty \, n \in A_{n_0}^{(j)}}} D\left(\frac{E_n^{(j)}}{(-1)^{j+1} h_n}, f'(x_0)\right) = 0.$$

**Example 8.35.** *Let $c \in \mathbb{R}_\mathcal{F}$ and $g : (a, b) \to \mathbb{R}$. If $g$ is differentiable on $x_0$ (in usual sense), then the function $f : (a, b) \to \mathbb{R}_\mathcal{F}$ defined by $f(x) = c \cdot g(x)$, $\forall x \in (a, b)$, is weakly generalized differentiable on $x_0$ and we have $f'(x_0) = c \cdot g'(x_0)$.*

*Let $f(t) = (1, 2, 3)e^{-t}$. Then it is easy to see that $f'(t) = -(1, 2, 3)e^{-t} = (-3e^{-t}, -2e^{-t}, -e^{-t})$.*

*Let $f(t) = (1, 2, 3) \sin t$, then $f'(t) = (1, 2, 3) \cos t$, $\forall t \in \mathbb{R}$.*

**Proposition 8.36.** *If $u(t) = (x(t), y(t), z(t))$ is triangular number valued function, then*

*a) If $u$ is (i)-differentiable (Hukuhara differentiable) then $u' = (x', y', z')$.*

*b) If $u$ is (ii)-differentiable then $u' = (z', y', x')$.*

**Proof.** The proof of b) is as follows. Let us suppose that the H-difference $u(t) \ominus u(t + h)$ exists. Then we get:

$$\lim_{h \searrow 0} \frac{u(t) \ominus u(t + h)}{-h}$$

$$= \lim_{h \searrow 0} \frac{(x(t) - x(t + h), y(t) - y(t + h), z(t) - z(t + h))}{-h}$$

$$= \lim_{h \searrow 0} \left(\frac{z(t) - z(t + h)}{-h}, \frac{y(t) - y(t + h)}{-h}, \frac{x(t) - x(t + h)}{-h}\right)$$

$$= (z', y', x').$$

Similarly

$$\lim_{h \searrow 0} \frac{u(t - h) \ominus u(t)}{-h} = (z', y', x'),$$

and the required conclusion follows.                                      ∎

Other generalizations were recently obtained.

**Definition 8.37.** (Stefanini-Bede [141], Bede-Stefanini [27]) Let $x_0 \in (a, b)$. Then the **fuzzy gH-derivative** of a function $f : (a, b) \to R_\mathcal{F}$ at $x_0$ is defined as

$$f'_{gH}(x_0) = \lim_{h \to 0} \frac{1}{h}[f(x_0 + h) \ominus_{gH} f(x_0)].$$

If $f'_{gH}(x_0) \in R_\mathcal{F}$ exists, then we say that $f$ is generalized Hukuhara differentiable (gH-differentiable for short) at $x_0$ ($\ominus_{gH}$ is the gH-difference) .

**Theorem 8.38.** *(Bede-Stefanini [27]) The gH-differentiability concept and the weakly generalized (Hukuhara) differentiability given in Definition 8.34 coincide.*

**Proof.** Let us suppose that $f$ is gH-differentiable (as in Definition 8.37). For any sequence $h_n \searrow 0$, for $n$ sufficiently large, at least two of the Hukuhara differences $f(x_0 + h_n) \ominus_H f(x_0)$, $f(x_0) \ominus_H f(x_0 + h_n)$, $f(x_0) \ominus_H f(x_0 - h_n)$, $f(x_0 - h_n) \ominus_H f(x_0)$ exist. As a conclusion we have

$$A_{n_0}^{(1)} \cup A_{n_0}^{(2)} \cup A_{n_0}^{(3)} \cup A_{n_0}^{(4)} = \{n \in N; n \geq n_0\}$$

for any $n_0 \in N$. Now we observe that

$$\frac{E_n^{(j)}}{(-1)^{j+1} h_n} = \frac{f(x_0 + h_n) \ominus_{gH} f(x_0)}{h_n},$$

so $f$ is weakly generalized differentiable.

Reciprocally, if we assume $f$ to be weakly generalized differentiable then since at least two of the sets $A_{n_0}^{(1)}, A_{n_0}^{(2)}, A_{n_0}^{(3)}, A_{n_0}^{(4)}$ are infinite we get

$$\lim_{h \to 0} \frac{1}{h}[f(x_0 + h) \ominus_{gH} f(x_0)] = \lim_{h_n \searrow 0} \frac{E_n^{(j)}}{(-1)^{j+1} h_n}$$

for at least two indices from $j \in \{1, 2, 3, 4\}$, so $f$ is gH-differentiable. As a conclusion weakly generalized (Hukuhara) differentiability is equivalent to gH-differentiability. ∎

**Theorem 8.39.** *(Bede-Stefanini [27]) Let $f : (a, b) \to \mathbb{R}_{\mathcal{F}}$ be such that $[f(x)]_\alpha = [f_\alpha^-(x), f_\alpha^+(x)]$. Suppose that the functions $f_\alpha^-(x)$ and $f_\alpha^+(x)$ are real-valued functions, differentiable with respect to $x$, uniformly in $\alpha \in [0, 1]$. Then the function $f(x)$ is gH-differentiable at a fixed $x \in ]a, b[$ if and only if one of the following two cases holds:*
*a) $(f_\alpha^-)'(x)$ is increasing, $(f_\alpha^+)'(x)$ is decreasing as functions of $\alpha$, and*

$$(f_1^-)'(x) \leq (f_1^+)'(x),$$

*or*
*b) $(f_\alpha^-)'(x)$ is decreasing, $(f_\alpha^+)'(x)$ is increasing as functions of $\alpha$, and*

$$(f_1^+)'(x) \leq (f_1^-)'(x).$$

*Moreover, $\forall \alpha \in [0, 1]$ we have*

$$\left[f_{gH}'(x)\right]_\alpha = [\min\{(f_\alpha^-)'(x), (f_\alpha^+)'(x)\}$$

$$, \max\{(f_\alpha^-)'(x), (f_\alpha^+)'(x)\}].$$

**Proof.** Let $f$ be gH-differentiable and assume that $f_\alpha^-(x)$ and $f_\alpha^+(x)$ are differentiable. Clearly, gH-differentiability implies that level-wise we have

$$[f'_{gH}(x)]_\alpha = [(f_\alpha^-)'(x), (f_\alpha^+)'(x)].$$

or

$$[f'_{gH}(x)]_\alpha = [(f_\alpha^+)'(x), (f_\alpha^-)'(x)].$$

In the first step let us suppose that there is a sign change in the difference of derivatives $(f_\alpha^+)'(x) - (f_\alpha^-)'(x)$ at a fixed $\alpha_0 \in (0, 1)$. Then $[f'_{gH}(x)]_{\alpha_0}$ is a singleton and, for all $\alpha$ such that $\alpha_0 \le \alpha \le 1$, also $[f'_{gH}(x)]_\alpha$ is a singleton because

$$[f'_{gH}(x)]_\alpha \subseteq [f'_{gH}(x)]_{\alpha_0}$$

it follows that, for the same values of $\alpha$,

$$(f_\alpha^+)'(x) - (f_\alpha^-)'(x) = 0$$

which is a contradiction with the fact that $(f_\alpha^+)'(x) - (f_\alpha^-)'(x)$ changes sign. We then conclude that $(f_\alpha^+)'(x) - (f_\alpha^-)'(x)$ cannot change sign.

Case 1. If $(f_1^-)'(x) < (f_1^+)'(x)$ then

$$(f_\alpha^+)'(x) - (f_\alpha^-)'(x) \ge 0$$

for every $\alpha \in [0, 1]$ and

$$[f'_{gH}(x)]_\alpha = [(f_\alpha^-)'(x), (f_\alpha^+)'(x)];$$

since $f$ is gH-differentiable, the intervals $[(f_\alpha^-)'(x), (f_\alpha^+)'(x)]$ should form a fuzzy number, i.e., for any $\alpha > \beta$,

$$[(f_\alpha^-)'(x), (f_\alpha^+)'(x)] \subseteq [(f_\beta^-)'(x), (f_\beta^+)'(x)]$$

which shows that $(f_\alpha^-)'(x)$ has to be increasing and $(f_\alpha^+)'(x)$ needs to be decreasing as a function of $\alpha$.

Case 2. If $(f_1^-)'(x) > (f_1^+)'(x)$, then

$$(f_\alpha^+)'(x) - (f_\alpha^-)'(x) \le 0$$

for every $\alpha \in [0, 1]$ and, in this case,

$$[f'_{gH}(x)]_\alpha = [(f_\alpha^+)'(x), (f_\alpha^-)'(x)]$$

so we get

$$[(f_\alpha^+)'(x), (f_\alpha^-)'(x)] \subseteq [(f_\beta^+)'(x), (f_\beta^-)'(x)],$$

for any $\alpha > \beta$, which shows that $(f_\alpha^-)'(x)$ is decreasing and $(f_\alpha^+)'(x)$ is increasing as a function of $\alpha$.

Case 3. If we have

$$(f_1^-)'(x) = (f_1^+)'(x)$$

and if $(f_{gH})'(x) \in \mathbb{R}$ is a crisp number, the conclusion is obvious. Otherwise we may have either

$$(f_0^-)'(x) < (f_0^+)'(x)$$

or

$$(f_0^-)'(x) > (f_0^+)'(x)$$

when $\alpha < \alpha_0$ and equality

$$(f_0^-)'(x) = (f_0^+)'(x)$$

for $\alpha \geq \alpha_0$. We have in this case the monotonicity satisfied.

Reciprocally, let us consider the Banach space $B = \bar{C}[0,1] \times \bar{C}[0,1]$, where $\bar{C}[0,1]$ is the space of left continuous functions on $(0,1]$, right continuous at $0$, with the uniform norm. For any fixed $x \in (a,b)$, the mapping $j_x : \mathbb{R}_{\mathcal{F}} \to B$, defined by

$$j_x(f) = (f^-(x), f^+(x)) = \{(f_\alpha^-(x), f_\alpha^+(x)) | \alpha \in [0,1]\},$$

is an isometric embedding. Assuming that, for all $\alpha$, the two functions $f_\alpha^-(x)$ and $f_\alpha^+(x)$ are differentiable with respect to $x$, the limits

$$(f_\alpha^-)'(x) = \lim_{h \to 0} \frac{f_\alpha^-(x+h) - f_\alpha^-(x)}{h}$$

$$(f_\alpha^+)'(x) = \lim_{h \to 0} \frac{f_\alpha^+(x+h) - f_\alpha^+(x)}{h}$$

exist uniformly for all $\alpha \in [0,1]$. Taking a sequence $h_n \to 0$, we will have

$$(f_\alpha^-)'(x) = \lim_{n \to \infty} \frac{f_\alpha^-(x+h_n) - f_\alpha^-(x)}{h_n}$$

$$(f_\alpha^+)'(x) = \lim_{n \to \infty} \frac{f_\alpha^+(x+h_n) - f_\alpha^+(x)}{h_n},$$

i.e., $(f_\alpha^-)'(x), (f_\alpha^+)'(x)$ are uniform limits of sequences of left continuous functions at $\alpha \in (0,1]$, so they are themselves left continuous for $\alpha \in (0,1]$. Similarly the right continuity at $0$ can obtained.

Assuming that, for a fixed $x \in [a,b]$, the function $(f_\alpha^-)'(x)$ is increasing and the function $(f_\alpha^+)'(x)$ is decreasing as functions of $\alpha$, and that

$$(f_1^-)'(x) \leq (f_1^+)'(x)$$

then also

$$(f_\alpha^-)'(x) \leq (f_\alpha^+)'(x), \forall \alpha \in [0,1]$$

and it is easy to see that the pair of functions $(f_\alpha^-)'(x)$, $(f_\alpha^+)'(x)$ fulfill the conditions in Theorem 4.8 and the intervals

$$[(f_\alpha^-)'(x), (f_\alpha^+)'(x)], \alpha \in [0, 1]$$

determine a fuzzy number. Now we observe that the following limit uniformly exists

$$\left[\lim_{h \to 0} \frac{f(x+h) \ominus_{gH} f(x)}{h}\right]_\alpha$$

$$= \left[\lim_{h \to 0} \frac{f_\alpha^-(x+h) - f_\alpha^-(x)}{h}, \lim_{h \to 0} \frac{f_\alpha^+(x+h) - f_\alpha^+(x)}{h}\right]$$

$$= [(f_\alpha^-)'(x), (f_\alpha^+)'(x)], \forall \alpha \in [0, 1],$$

and it is a fuzzy number. As a conclusion, finally we obtain that $f$ is gH-differentiable. The symmetric case being analogous, the proof is complete.                                                                                        ∎

**Remark 8.40.** *Any of the derivative concepts defined above can be restricted to the case of closed real intervals. For example $f : (a, b) \to \mathbb{I}$, with*

$$\mathbb{I} = \{[a, b] | a \le b, a, b \in \mathbb{R}\}$$

*has its gH-derivative (Stefanini-Bede [141])*

$$f'_{gH}(x) = \lim_{n \to \infty} \frac{f(x+h) \ominus_{gH} f(x)}{h}$$

$$= [\min\{(f^-)'(x), (f^+)'(x)\}, \max\{(f^-)'(x), (f^+)'(x)\}].$$

*This expression always represents a well-defined real interval.*

Based on the g-difference, we consider the following g-differentiability concept, that further extends the gH-differentiability.

**Definition 8.41.** (Stefanini-Bede [27]) Let $x_0 \in (a, b)$. Then the **fuzzy g-derivative** of a function $f : (a, b) \to R_\mathcal{F}$ at $x_0$ is defined as

$$f'_g(x_0) = \lim_{h \to 0} \frac{1}{h}[f(x_0 + h) \ominus_g f(x_0)].$$

If $f'_g(x_0) \in \mathbb{R}_\mathcal{F}$ exists, we say that $f$ is generalized differentiable (g-differentiable for short) at $x_0$.

**Remark 8.42.** *We observe that the g-derivative is the most general among the previous definitions. Indeed,*

$$f(x_0 + h) \ominus_g f(x_0) = f(x_0 + h) \ominus_{gH} f(x_0)$$

*whenever the gH-difference exists. This also implies that $f'_g(x) = f'_{gH}(x)$ whenever this later exists.*

In the following Theorem we prove a characterization and a practical formula for the g-derivative.

**Theorem 8.43.** *(Stefanini-Bede [27]) Let $f : [a, b] \to \mathbb{R}_\mathcal{F}$ be such that*

$$[f(x)]_\alpha = [f_\alpha^-(x), f_\alpha^+(x)], \alpha \in [0, 1].$$

*If $f_\alpha^-(x)$ and $f_\alpha^+(x)$ are differentiable real-valued functions with respect to $x$, uniformly with respect to $\alpha \in [0, 1]$, then $f(x)$ is g-differentiable and we have*

$$[f_g'(x)]_\alpha = \left[\inf_{\beta \geq \alpha} \min\{(f_\beta^-)'(x), (f_\beta^+)'(x)\}, \sup_{\beta \geq \alpha} \max\{(f_\beta^-)', (f_\beta^+)'(x)\}\right].$$

**Proof.** By Proposition 5.19 we have

$$\frac{1}{h}[f(x+h) \ominus_g f(x)]_\alpha$$

$$= \frac{1}{h}[\inf_{\beta \geq \alpha} \min\{f(x+h)_\beta^- - f(x)_\beta^-, f(x+h)_\beta^+ - f(x)_\beta^+\},$$

$$\sup_{\beta \geq \alpha} \max\{f(x+h)_\beta^- - f(x)_\beta^-, f(x+h)_\beta^+ - f(x)_\beta^+\}].$$

Since $f_\alpha^-(x), f_\alpha^+(x)$ are differentiable we obtain

$$\lim_{h \to 0} \frac{1}{h}[f(x+h) \ominus_g f(x)]_\alpha$$

$$= \left[\inf_{\beta \geq \alpha} \min\{(f_\beta^-)'(x), (f_\beta^+)'(x)\}, \sup_{\beta \geq \alpha} \max\{(f_\beta^-)'(x), (f_\beta^+)'(x)\}\right]$$

for any $\alpha \in [0, 1]$. Also, let us observe that if $f_\alpha^-, f_\alpha^+$ are left continuous with respect to $\alpha \in (0, 1]$ and right continuous at 0, considering a sequence $h_n \to 0$, the functions

$$\frac{f_\alpha^-(x+h_n) - f_\alpha^-(x)}{h_n}, \frac{f_\alpha^+(x+h_n) - f_\alpha^+(x)}{h_n}$$

are left continuous at $\alpha \in (0, 1]$ and right continuous at 0. Also, the functions

$$\inf_{\beta \geq \alpha} \min \left\{ \frac{f_\beta^-(x+h_n) - f_\beta^-(x)}{h_n}, \frac{f_\beta^+(x+h_n) - f_\beta^+(x)}{h_n} \right\}$$

and

$$\sup_{\beta \geq \alpha} \max \left\{ \frac{f_\beta^-(x+h_n) - f_\beta^-(x)}{h_n}, \frac{f_\beta^+(x+h_n) - f_\beta^+(x)}{h_n} \right\}$$

fulfill the same properties. A limit of a sequence of left (right) continuous functions is left (right) continuous. Then it follows that

$$\inf_{\beta \geq \alpha} \min\{\left(f_\beta^-\right)'(x), \left(f_\beta^+\right)'(x)\},$$

$$\sup_{\beta \geq \alpha} \max\{\left(f_\beta^-\right)'(x), \left(f_\beta^+\right)'(x)\}$$

are left continuous for $\alpha \in (0,1]$ and right continuous at 0. To check monotonicity properties we observe that for $\alpha_1 \leq \alpha_2$ we have

$$\{\left(f_\beta^-\right)'(x), \left(f_\beta^+\right)'(x)|\beta \geq \alpha_2\} \subseteq \{\left(f_\beta^-\right)'(x), \left(f_\beta^+\right)'(x)|\beta \geq \alpha_1\}.$$

Then

$$\inf_{\beta \geq \alpha} \min\{\left(f_\beta^-\right)'(x), \left(f_\beta^+\right)'(x)\}$$

is increasing with respect to $\alpha \in [0,1]$ and

$$\sup_{\beta \geq \alpha} \max\{\left(f_\beta^-\right)'(x), \left(f_\beta^+\right)'(x)\}$$

is decreasing in $\alpha \in [0,1]$. The condition

$$\min\{\left(f_1^-\right)'(x), \left(f_1^+\right)'(x)\} \leq \max\{\left(f_1^-\right)'(x), \left(f_1^+\right)'(x)\}$$

concludes the reasoning that allows us to infer that the above functions define a fuzzy number with

$$[f_g'(x)]_\alpha = \left[\inf_{\beta \geq \alpha} \min\{\left(f_\beta^-\right)'(x), \left(f_\beta^+\right)'(x)\}, \sup_{\beta \geq \alpha} \max\{\left(f_\beta^-\right)', \left(f_\beta^+\right)'(x)\}\right].$$

As a conclusion, the level sets $[f_g'(x)]_\alpha$ define a fuzzy number, and so, the derivative $f_g'(x)$ exists in the sense of the g-derivative.  ∎

**Theorem 8.44.** *(Bede-Stefanini [27]) Let $f : (a,b) \to \mathbb{R}_\mathcal{F}$ be uniformly level-wise gH-differentiable at $x_0$. Then $f$ is g-differentiable at $x_0$ and, for any $\alpha \in [0,1]$,*

$$[f_g'(x_0)]_\alpha = cl\left(\bigcup_{\beta \geq \alpha} (f_{gH}'(x_0))_\beta\right)$$

**Proof.** For the definition of the g-difference (Definition 5.18) we have

$$[f(x_0 + h) \ominus_g f(x_0)]_\alpha = cl \bigcup_{\beta \geq \alpha} ([f(x_0 + h)]_\beta \ominus_{gH} [f(x_0)]_\beta), \forall \alpha \in [0, 1].$$

Taking the uniform limit in the above relation we obtain

$$[f'_g(x_0)]_\alpha = cl \left( \bigcup_{\beta \geq \alpha} (f'_{gH}(x_0))_\beta \right).$$

∎

We conclude with an example of a fuzzy valued function that is not gH-differentiable but it is g-differentiable.

**Example 8.45.** *Consider the fuzzy valued function* $f : [-2, 2] \to \mathbb{R}_\mathcal{F}$ *having triangular values as outputs:*

$$f(x) = \left( \frac{x^3}{3}, \frac{x^3}{3} + x + 3, \frac{2x^3}{3} + 4 \right)$$

*having the level sets defined by*

$$f_\alpha^-(x) = \frac{x^3}{3} + \alpha(x + 3)$$

$$f_\alpha^+(x) = (2 - \alpha)\frac{x^3}{3} + x\alpha + 4 - \alpha.$$

*We have*

$$\left( f_\alpha^- \right)'(x) = x^2 + \alpha$$

$$\left( f_\alpha^+ \right)'(x) = (2 - \alpha)x^2 + \alpha.$$

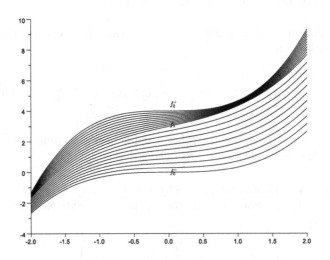

**Fig. 8.1** The level sets of a fuzzy valued function

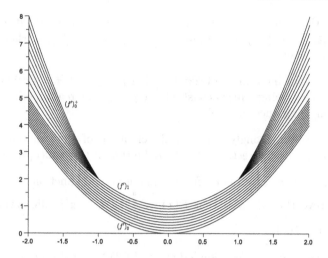

**Fig. 8.2** The level sets of the g-derivative of a fuzzy valued function

*We observe that on the interval $[-2, -1]$ and $[1, 2]$ intervals the function is gH-differentiable, namely it is Hukuhara differentiable. In the interval $[-1, 1]$ it is not gH-differentiable but it is g-differentiable. The function and its g-derivative are represented in Figs. 8.1 and 8.2 respectively.*

## 8.10   Problems

1. Calculate the Haussdorff distance between two arbitrary triangular fuzzy numbers $u = (a_1, b_1, c_1)$ and $v = (a_2, b_2, c_2)$ in a simple form. Can we extend this result to trapezoidal fuzzy numbers? (Numerical examples $u = (-1, 1, 2)$, $v = (3, 4, 6)$ and respectively $u = (1, 2, 3, 5)$, $v = (3, 3, 4, 6)$.

2. Calculate the $D_1$ distance between two arbitrary triangular fuzzy numbers $u = (a_1, b_1, c_1)$ and $v = (a_2, b_2, c_2)$ in a simple form. Can we extend this result to trapezoidal fuzzy numbers? (Numerical examples $u = (-1, 1, 2)$, $v = (3, 4, 6)$ and respectively $u = (1, 2, 3, 5)$, $v = (3, 3, 4, 6)$.

3. Prove the properties of the $D_p$ distance listed in Proposition 8.4.

4. Prove the properties (i)-(iv) and (vi) of a norm of a fuzzy number in Proposition 8.13.

5. Show that the Hausdorff distance is linked to the generalized difference through the relation

$$D_\infty(u, v) = \|u \ominus_g v\|_{\mathcal{F}}.$$

6. Prove that the space of continuous fuzzy-number-valued functions $(C\left([a,b]\,;\mathbb{R}_{\mathcal{F}}\right),D_{\infty})$ is a complete metric space and also, that it is an FN-type space.

7. Calculate the integral of a fuzzy function $f : [a,b] \to \mathbb{R}_{\mathcal{F}}$, $f(x) = c \cdot g(x)$, with $c \in \mathbb{R}_{\mathcal{F}}$ being a fuzzy constant and $g(x) \in \mathbb{R}$ a real function possibly changing sign once in $(a,b)$.

8. Calculate the strongly generalized derivative of the function $f(x) = (1,2,3,5) \cdot \sin x$, directly using the definition of this concept.

9. a) For a given continuous fuzzy-number-valued function $f : [a,b] \to \mathbb{R}_{\mathcal{F}}$, prove that the function $F(x) = \int\limits_{a}^{x} f(t)dt$ is gH-differentiable and $F'_{gH}(x) = f(x)$,

   b) Prove that the function $G(x) = \int\limits_{x}^{b} f(t)dt$ is gH-differentiable and $G'_{gH}(x) = -f(x)$.

10. Prove that if $f$ is a strongly generalized differentiable function as in case (i) or (ii) of strongly generalized differentiability over the entire interval $[a,b]$ then we have

$$\int_{a}^{b} f'_{gH}(x)dx = f(b) \ominus_{gH} f(a).$$

11. Prove that the function given level-wise for $\alpha \in [0,1]$ as

$$f_{\alpha}^{-}(x) = xe^{-x} + \alpha^2 \left( e^{-x^2} + x - xe^{-x} \right)$$
$$f_{\alpha}^{+}(x) = e^{-x^2} + x + (1 - \alpha^2) \left( e^x - x + e^{-x^2} \right)$$

is g-differentiable without being gH-differentiable.

# 9
# Fuzzy Differential Equations

Fuzzy differential equations (FDEs) appear as a natural way to model the propagation of epistemic uncertainty in a dynamical environment. There are several interpretations of a fuzzy differential equation. The first one historically was based on the Hukuhara derivative introduced in Puri-Ralescu [123] and studied in several papers (Wu-Song-Lee [150], Kaleva [83], Ding-Ma-Kandel [46], Rodriguez-Lopez [125]). This interpretation has the disadvantage that solutions of a fuzzy differential equation have always an increasing length of the support. This fact implies that the future behavior of a fuzzy dynamical system is more and more uncertain in time. This phenomenon does not allow the existence of periodic solutions or asymptotic phenomena. That is why different ideas and methods to solve fuzzy differential equations have been developed. One of them solves differential equations using Zadeh's extension principle (Buckley-Feuring [30]), while another approach interprets fuzzy differential equations through differential inclusions. Differential inclusions and Fuzzy Differential Inclusions are two topics that are very interesting but they do not constitute the subject of the present work (see Diamond [45], Lakshmikantham-Mohapatra [98]). Recently new approaches have been developed based on generalized fuzzy derivatives discussed in the previous chapter. In the present work we will work with the interpretations based on Hukuhara differentiability, Zadeh's extension principle and the strongly generalized differentiability concepts.

## 9.1 FDEs under Hukuhara Differentiability

In this section we consider the fuzzy initial value problem $x' = f(t, x)$, $x(t_0) = x_0 \in \mathbb{R}_{\mathcal{F}}$ under Hukuhara differentiability. The function $f : \mathbb{R} \times \mathbb{R}_{\mathcal{F}} \to \mathbb{R}_{\mathcal{F}}$

B. Bede: *Mathematics of Fuzzy Sets and Fuzzy Logic*, STUDFUZZ 295, pp. 171–191.
DOI: 10.1007/978-3-642-35221-8_9      © Springer-Verlag Berlin Heidelberg 2013

is assumed to be continuous. The following lemma transforms the fuzzy differential equation into integral equations. In the present section the differentiability concept is that of Hukuhara differentiability. A fuzzy differential equation is written using the Hukuhara derivative.

**Lemma 9.1.** *For $t_0 \in \mathbb{R}$, the fuzzy differential equation $x' = f(t, x)$, $x(t_0) = x_0 \in \mathbb{R}_{\mathcal{F}}$ where $f : \mathbb{R} \times \mathbb{R}_{\mathcal{F}} \to \mathbb{R}_{\mathcal{F}}$ is continuous, is equivalent to the integral equation*

$$x(t) = x_0 + \int_{t_0}^t f(s, x(s))ds,$$

*on some interval $[t_0, t_1] \subset \mathbb{R}$.*

**Proof.** Let us suppose that $x$ is a solution of the differential equation $x' = f(t, x)$, $x(t_0) = x_0 \in \mathbb{R}_{\mathcal{F}}$. Then by integration we get

$$\int_{t_0}^t x'(s)ds = \int_{t_0}^t f(s, x(s))ds$$

and using the result in problem 10 of the previous chapter we get

$$x(t) \ominus x_0 = \int_{t_0}^t f(s, x(s))ds,$$

i.e.,

$$x(t) = x_0 + \int_{t_0}^t f(s, x(s))ds.$$

Reciprocally given a solution $x$ of the integral equation we can write

$$x(t + h) = x_0 + \int_{t_0}^{t+h} f(s, x(s))ds$$

and

$$\lim_{h \searrow 0} \frac{x(t + h) \ominus x(t)}{h} = \lim_{h \searrow 0} \frac{1}{h} \int_t^{t+h} f(s, x(s))ds.$$

We observe that

$$D \left[ \int_t^{t+h} f(s, x(s))ds, h \cdot f(t, x(t)) \right]$$

$$= D \left[ \int_t^{t+h} f(s, x(s))ds, \int_t^{t+h} f(t, x(t))ds \right]$$

$$\leq \int_t^{t+h} D(f(s, x(s)), f(t, x(t)))ds$$

$$\leq \int_t^{t+h} \omega(f(t, x(t)), h)ds = h \cdot \omega(f(t, x(t)), h),$$

where $\omega(f(t, x(t)), h)$ denotes the modulus of continuity of the function $f(t, x(t))$ which is a continuous as a function of $t \in [t_0, t_1]$. Then

$$\lim_{h \searrow 0} D\left(\frac{x(t+h) \ominus x(t)}{h}, f(t, x(t))\right)$$

$$= \lim_{h \searrow 0} \frac{1}{h} h \cdot \omega(f(t, x(t)), h) = 0$$

and this implies that $x$ is a solution of the fuzzy initial value problem $x' = f(t, x), x(t_0) = x_0$. ∎

In Song-Wu-Lee [150] the existence of uniqueness and solutions in this interpretation was proved. This approach is close to Pickard-Lindelöf Theorem and uses Banach fixed-point Theorem in the proof. Different approaches based on Arzela-Ascoli results are found in literature, but they usually fail on the extraction of a convergent subsequence of a bounded sequence.

First we will show that Lipshitz functions are also bounded, following Lupulescu.

**Lemma 9.2.** ( Lupulescu [103]) Let $R_0 = [t_0, t_0 + p] \times \overline{B}(x_0, q)$ and let us assume that $f : R_0 \to \mathbb{R}_{\mathcal{F}}$ is continuous and fulfills the Lipschitz condition

$$D(f(t, x), f(t, y)) \leq L \cdot D(x, y), \forall (t, x), (t, y) \in R_0.$$

Then $f$ is bounded, i.e., there exists $M > 0$ such that

$$D(f(t, x), 0) \leq M.$$

**Proof.** We observe that

$$D(f(t, x), 0) \leq D(f(t, x), f(t, x_0)) + D(f(t, x_0), 0)$$

The real function $D(f(t, x_0), 0)$ is bounded when $t \in [t_0, t_0 + p]$, i.e., there exists $M_1$ with $D(f(t, x_0), 0) \leq M_1$. We get

$$D(f(t, x), 0) \leq L \cdot D(x, x_0) + M_1 \leq Lq + M_1 = M,$$

i.e., $f$ is bounded. ∎

**Theorem 9.3.** (Wu-Song-Lee [150], Lupulescu [103], Bede-Gal [21]) Let $R_0 = [t_0, t_0 + p] \times \overline{B}(x_0, p), p > 0, x_0 \in \mathbb{R}_{\mathcal{F}}$ and $f : R_0 \to \mathbb{R}_{\mathcal{F}}$ be continuous such that the following Lipschitz condition holds:
There exist a constant $L > 0$ such that

$$D(f(t, x), f(t, y)) \leq L \cdot D(x, y), \forall (t, x), (t, y) \in R_0.$$

Then the Fuzzy Initial Value Problem (FIVP)

$$x'(t) = f(t, x), \quad x(t_0) = x_0,$$

has a unique solution defined in an interval $[t_0, t_0 + k]$ for some $k > 0$.

**Proof.** Let us consider $K_0 = C([t_0, t_0+p], \mathbb{R}_{\mathcal{F}})$ and the operator $P : K_0 \to K_0$ defined by:

$$P(x_0)(t) = x_0,$$

$$P(x)(t) = x_0 + \int_{t_0}^{t} f(s, x(s))ds$$

It is easy to see that $P$ is well defined. By Lemma 9.2, and the Lipschitz condition, $f$ is bounded and also $P$ is bounded,

$$D(P(x)(t), x_0) \le \int_{t_0}^{t} D(f(s, x(s)), 0)ds \le M(t - t_0)$$

where

$$M = \sup_{(t,x) \in R_0} D(f(t, x), 0)$$

is provided by Lemma 9.2. Let $d = \min\{p, \frac{q}{M}\}$ and

$$K_1 = C([t_0, t_0 + d], \overline{B}(x_0, q)).$$

Let us consider now $P : K_1 \to C([t_0, t_0 + d], \mathbb{R}_{\mathcal{F}})$. We have

$$D(P(x)(t), x_0) \le q$$

and so given $x \in K_1$ gives $P(x) \in K_1$. We observe that $K_1$ is a complete metric space considered with the uniform distance, as a closed subspace of a complete metric space.

Now we prove that $P$ is a contraction. Indeed,

$$D(P(x)(t), P(y)(t)) \le \int_{t_0}^{t} D(f(s, x(s)), f(s, y(s)))ds$$

$$\le 2L(t - t_0)D(x, y).$$

Now choosing $k = \min\{d, \frac{1}{2L}\}$ and further restricting $P : K_2 \to K_2$, with

$$K_2 = C([t_0, t_0 + k], \overline{B}(x_0, q))$$

we obtain that $P$ is a contraction. From Banach's fixed point theorem there exists a fixed point $x_* \in K_2$ with $P(x_*) = x_*$ i.e., $x_*$ is a solution of Lemma 9.1. Finally from Lemma 9.1 we obtain that $x_*$ is a solution of the FIVP

$$x'(t) = f(t, x(t)), \quad x(t_0) = x_0,$$

for $t \in [t_0, t_0 + k]$. The uniqueness follows from the uniqueness of the fixed point of $P$, which is a consequence of Banach's fixed point Theorem.    ∎

Next we prove a characterization result.

**Theorem 9.4.** *(Bede [15]) Let $R_0 = [t_0, t_0 + p] \times \overline{B}(x_0, p)$, $p > 0$, $x_0 \in \mathbb{R}_{\mathcal{F}}$ and $f : R_0 \to \mathbb{R}_{\mathcal{F}}$ be continuous such that*

$$f(t, x)_r = [f_r^-(t, x_r^-, x_r^+), f_r^+(t, x_r^-, x_r^+)], r \in [0, 1].$$

*If*

$$f_r^-(t, x_r^-, x_r^+), f_r^+(t, x_r^-, x_r^+), r \in [0, 1]$$

*are equicontinuous ($\forall \varepsilon > 0, \exists \delta > 0$ such that*

$$|f_r^-(t, x_r^-, x_r^+) - f_r^-(t_0, (x_0)_r^-, (x_0)_r^+)| < \varepsilon$$

*when*

$$\left\| (t, x_r^-, x_r^+) - (t_0, (x_0)_r^-, (x_0)_r^+) \right\| < \delta,$$

$\forall r \in [0, 1]$) *and uniformly Lipschitz in the second and third argument i.e., there exist a constant $L > 0$ such that*

$$\left| f_r^\pm(t, x_r^-, x_r^+), -f(t, y_r^-, y_r^+) \right| \leq L \cdot (|x_r^- - y_r^-| + |x_r^+ - y_r^+|),$$

*for any $(t, x), (t, y) \in R_0$ and for any $r \in [0, 1]$. Then the Fuzzy Initial Value Problem (FIVP)*

$$x'(t) = f(t, x), \ x(t_0) = x_0,$$

*has a unique solution defined in an interval $[t_0, t_0 + k]$ for some $k > 0$ and moreover, the unique solution is level-wise $x_r = [x_r^-, x_r^+]$ characterized by the system of ODEs*

$$\begin{cases} (x_r^-)' = f_r^-(t, x_r^-, x_r^+) \\ (x_r^+)' = f_r^+(t, x_r^-, x_r^+) \end{cases}, r \in [0, 1].$$

**Proof.** It is easy to see that the conditions of the theorem ensure by the classical Picard-Lindelöf Theorem the existence and uniqueness of solutions for the given crisp system. Also, equicontinuity ensures that $f(t, x)$ is continuous as a fuzzy number valued function. The uniform Lipschitz condition provides Lipschitz condition in Theorem 9.3 and so the existence of the unique solution of the fuzzy initial value problem

$$x' = f(t, x), x(t_0) = x_0.$$

Now, a Hukuhara differentiable function $x(t)$ has level sets differentiable and

$$(x')_r = [(x_r^-)', (x_r^+)'],$$

$r \in [0, 1]$. Also, the equation $x' = f(t, x)$ level-wise holds so we get

$$\begin{cases} (x_r^-)' = f_r^-(t, x_r^-, x_r^+) \\ (x_r^+)' = f_r^+(t, x_r^-, x_r^+) \end{cases}, r \in [0, 1].$$

But this equation has a unique solution. As a conclusion we obtain that the unique solution of the system characterizes the unique solution of the fuzzy initial value problem. ∎

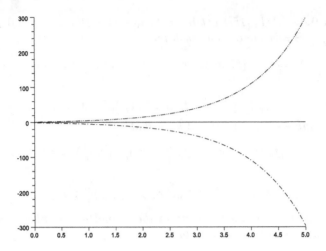

**Fig. 9.1** Hukuhara solution of a fuzzy differential equation

**Example 9.5.** *Consider as an example the FIVP*

$$x' = -x + 2e^{-t}(-1,0,1), \quad x(0) = (-1,0,1).$$

*It is easy to check that the conditions in Theorem 9.3 and also 9.4 are met. Also, based on Proposition 8.36. We search for triangular solution of the form* $x = (x_0^-, x_1, x_0^+)$. *Then we have*

$$\begin{cases} \left(x_0^-\right)' = -x_0^+ - 2e^{-t} \\ \quad x_1' = -x_1 \\ \left(x_0^+\right)' = -x_0^- + 2e^{-t} \end{cases}$$

*and we obtain the solution*

$$x(t) = (e^{-t} - 2e^t, 0, 2e^t - e^{-t}), \quad t \in (0, \infty)$$

*represented in Fig. 9.1*

## 9.2    The Interpretation Based on Zadeh's Extension Principle

Under this interpretation, an FIVP is solved as follows: We consider the crisp ODE which leads to the fuzzy equation considered and solve it. Then, solution of the FIVP are generated by using Zadeh's extension principle on the classical solution. So, we start with a classical ODE $x' = f(t, x, a)$, $x(t_0) = x_0 \in \mathbb{R}$, where $a \in \mathbb{R}$ is a parameter that appears in the given differential equation. Let us recall the following result concerning existence, uniqueness and continuous dependence on the parameters and initial value of an ODE.

**Theorem 9.6.** *(see e.g. Perko [122]) Let*

$$f : [t_0, t_0 + p] \times [x_0 - q, x_0 + q] \times \bar{B}(a_0, r) \to \mathbb{R}$$

*(here*

$$\bar{B}_{\mathbb{R}^n}(a_0, r) = \{a \in \mathbb{R}^n | \, \|a - a_0\| \leq r\} \subseteq \mathbb{R}^n$$

*denotes a closed ball in $\mathbb{R}^n$ with the Euclidean norm $\|\cdot\|$). Let us assume that $f$ is Lipschitz in its second variable i.e. there exists $L_1$ such that,*

$$\|f(t, x, a) - f(t, y, a)\| \leq L_1 \, \|x - y\| \, .$$

*Let us further assume that $f$ is Lipschitz in the third variable i.e., there exists $L_2$ such that,*

$$\|f(t, x, a) - f(t, x, b)\| \leq L_2 \, \|a - b\| \, .$$

*Then the initial value problem*

$$x' = f(t, x, a), x(t_0) = x_0,$$

*has a unique solution. Moreover, the unique solution depends continuously both on the initial condition and parameters.*

**Proof.** The proof can be found in e.g., Perko [122] (we omit this proof since it is not involving fuzzy sets) ∎

Let $\tilde{f} : [t_0, t_0 + p] \times \bar{B}(x_0, q) \times \bar{B}(a_0, r) \to \mathbb{R}_{\mathcal{F}}$ (here $\bar{B}(x_0, q), \bar{B}(a_0, r)$ are closed balls in $\mathbb{R}_{\mathcal{F}}$) be the Zadeh extension of $f(t, x, a)$.

**Definition 9.7.** *Let $A, X_0 \in \mathbb{R}_{\mathcal{F}}$ be fuzzy numbers and $\tilde{f}$ the Zadeh extension of $f$. The fuzzy initial value problem*

$$X' = \tilde{f}(t, X, A), X(t_0) = X_0,$$

*is said to have the solution $X : [t_0, t_0 + p] \to \mathbb{R}_{\mathcal{F}}$ where $X$ is the Zadeh extension of the solution $x : [t_0, t_0 + p] \to \mathbb{R}$ of the classical problem $x' = f(t, x, a), x(t_0) = x_0$.*

Based on the previous Theorem combined with Theorem 5.6 we can obtain the following existence and uniqueness result.

**Theorem 9.8.** *Let $f : [t_0, t_0 + p] \times [x_0 - q, x_0 + q] \times \bar{B}(a_0, r)$. Let us assume that $f$ is Lipschitz in its second variable i.e. there exists $L_1$ such that,*

$$\|f(t, x, a) - f(t, y, a)\| \leq L_1 \, \|x - y\|$$

*and that $f$ is Lipschitz in the third variable i.e., there exists $L_2$ such that,*

$$\|f(t, x, a) - f(t, x, b)\| \leq L_2 \, \|a - b\| \, .$$

*Then the solution of the initial value problem*

$$X' = \tilde{f}(t, X, A), X(t_0) = X_0,$$

*interpreted as in Definition 9.7 is well defined and continuous. Moreover $X$ can be defined level-wise as*

$$X_r = x\left(t, (X_0)_r, A_r\right) = \{x(t, x_0, a) | x_0 \in (X_0)_r, a \in A_r\},$$

*where $x(t, x_0, a)$ denotes the unique solution of the*

$$x' = f(t, x, a), x(t_0) = x_0$$

*classical problem.*

**Proof.** According to Theorem 9.6, the problem

$$x' = f(t, x, a), x(t_0) = x_0$$

has a unique solution $x(t, x_0, a)$ that depends continuously on $x_0, a$. From Theorem 5.6 the Zadeh extension of $x(t, x_0, a)$ is unique, well defined and continuous. Let us denote it by $X(t, X_0 A)$. Also, from Theorem 5.6 we get level-wise for $r \in [0, 1]$ that

$$X_r = x\left(t, (X_0)_r, A_r\right) = \{x(t, x_0, a) | x_0 \in X_0, a \in A\}.$$

∎

Let us consider the following very simple FIVP

$$\begin{cases} x' = -(1, 2, 3) \cdot x \\ x(0) = (1, 2, 3) \end{cases}.$$

with triangular fuzzy numbers. We consider this problem under the interpretation using the extension principle.

The solution of the problem obtained symbolically from the IVP for the ODE

$$\begin{cases} x' = -a \cdot x \\ x(0) = x_0 \end{cases}$$

with the solution

$$x(t) = x_0 e^{-at}.$$

Using Zadeh's extension principle we obtain the fuzzy solution

$$x(t) = (1, 2, 3)e^{-(1,2,3)t},$$

which can be written level-wise as

$$x(t)_r^- = (1 + r)e^{-(3-r)t}$$

$$x(t)_r^- = (3 - r)e^{-(1+r)t}.$$

which exists for $t \in [0, 1]$.

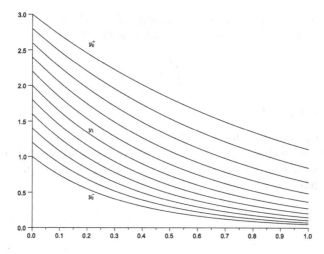

**Fig. 9.2** Zadeh extension based solution of a fuzzy differential equation

## 9.3  Fuzzy Differential Equations under Strongly Generalized Differentiability

### 9.3.1  Existence and Uniqueness of Two Solutions

Fuzzy differential equations under generalized differentiability were investigated first in Bede-Gal [20]. Later more general results were proposed in Bede-Gal [21]. Under strongly generalized differentiability there is a new behavior that emerges in the theory of differential equations in abstract space, namely that of the existence and local uniqueness of two solutions and that of switching. These very interesting phenomena are under investigation by many authors (Bede-Rudas-Gal [25], Chalco-Cano-Roman-Flores [36], Malinowski [106], Lupulescu [103] etc.) Other interesting research topics include the topics of Semigroups of operators on FN-type spaces (Gal-Gal [71], Gal-Gal-Guerekata [72], Kaleva [85])

We need one lemma:

**Lemma 9.9.** *(Bede-Gal, [21]) Let $x \in \mathbb{R}_{\mathcal{F}}$ be such that the functions $x_r = [x_r^-, x_r^+]$, $r \in [0,1]$ are differentiable, with $x^-$ strictly increasing and $x^+$ strictly decreasing on $[0,1]$, such that there exist the constants $c_1 > 0$, $c_2 < 0$ satisfying $(x_\alpha^-)' \geq c_1$ and $(x_\alpha^+)' \leq c_2$ for all $\alpha \in [0,1]$.*

*Let $f : [a,b] \to \mathbb{R}_{\mathcal{F}}$ be continuous with respect to $t$, having the level sets $f_\alpha^-(t)$ and $f_\alpha^+(t)$ with bounded partial derivatives $\frac{\partial f_\alpha^-(t)}{\partial \alpha}$ and $\frac{\partial f_\alpha^+(t)}{\partial \alpha}$, with respect to $\alpha \in [0,1]$, $t \in [a,b]$.*

*If*
*a) $x_1^- < x_1^+$*
*or if*

*b) $x_1^- = x_1^+$ and the core $[f(s)]_1$ consists of exactly one element for any $s \in T = [a, b]$, then there exists $h > a$ such that the H-difference*

$$x \ominus \int_a^t f(s)ds$$

*exists for any $t \in [a, h]$.*

**Proof.** We observe that the H-difference of two fuzzy numbers $u \ominus v$ exists if and only if the functions $(u^- - v^-, u^+ - v^+)$ define a fuzzy number. Indeed, let us suppose that $u \ominus v = z$, this is equivalent to $z + v = u$ and taking the $\alpha$-levels we have equivalently

$$[z_\alpha^-, z_\alpha^+] + [v_\alpha^-, v_\alpha^+] = [u_\alpha^-, u_\alpha^+],$$

i.e.,

$$z_\alpha^- = u_\alpha^- - v_\alpha^-,$$
$$z_\alpha^+ = u_\alpha^+ - v_\alpha^+,$$

$\alpha \in [0, 1]$. By Theorem 4.10, the assumption that $u \ominus v$ exists is equivalent with the fact that $(u^- - v^-, u^+ - v^+)$ define a fuzzy number. Moreover, since left and right continuity requirements of Theorem 4.10 obviously hold true whenever $u, v$ are fuzzy numbers, the existence of $u \ominus v$ becomes equivalent to

$$v_1^+ - v_1^- \le u_1^+ - u_1^-,$$

and that $u^- - v^-$ is non-decreasing and $u^+ - v^+$ is non-increasing.

Therefore, in order to prove the existence of $x \ominus \int_a^t f(s)ds$ in the statement, we have to check that

$$\left[\int_a^t f(s)ds\right]_1^+ - \left[\int_a^t f(s)ds\right]_1^- \le x_1^+ - x_1^- = len(x_1),$$

$$x_\alpha^- - \left[\int_a^t f(s)ds\right]_\alpha^- \text{ is nondecreasing w.r.t. } \alpha,$$

$$x_\alpha^+ - \left[\int_a^t f(s)ds\right]_\alpha^+ \text{ is nonincreasing w.r.t. } \alpha.$$

But

$$\left[\int_a^t f(s)ds\right]_\alpha = \left[\int_a^t f_\alpha^-(s)ds, \int_a^t f_\alpha^+(s)ds\right],$$

which implies that the above conditions are equivalent to

$$\int_a^t len([f(s)]_1)ds \le len([x]_1),$$

$$(x_\alpha^-)' - \int_a^t \frac{\partial f_\alpha^-(s)}{\partial \alpha} ds \geq 0, \text{ for all } \alpha \in [0,1]$$

and

$$(x_\alpha^+)' - \int_a^t \frac{\partial f_\alpha^+(s)}{\partial \alpha} ds \leq 0, \text{ for all } \alpha \in [0,1].$$

Since $f$ is continuous, it is bounded and the function $len[f(t)]_1$ is bounded as well. Let $M$ be such that

$$len[f(t)]_1 \leq M, t \in [a,b].$$

Also, note that we always have

$$\int_a^t len[f(s)]_1 ds \leq M(t-a).$$

Suppose we are under assumption a) in statement. Then, since for all $t \in [a, a + len([x]_1)/M]$, we get

$$M(t-a) \leq len([x]_1),$$

by the above inequality it easily follows that

$$\int_a^t len([f(s)]_1) ds \leq len([x]_1).$$

Let $M_1, M_2 > 0$ be such that

$$\left| \frac{\partial f_\alpha^-(s)}{\partial \alpha} \right| \leq M_1$$

and

$$\left| \frac{\partial f_\alpha^-(s)}{\partial \alpha} \right| \leq M_2,$$

for all $s \in [a,b]$ and $\alpha \in [0,1]$. Since $(x_\alpha^-)' \geq c_1$ for all $\alpha \in [0,1]$, we have

$$\int_a^t \frac{\partial f_\alpha^-(s)}{\partial \alpha} ds \leq (t-a)M_1 \leq c_1 \leq (x_\alpha^-)'$$

for any $t \in \left[a, a + \frac{c_1}{M_1}\right]$ and for all $\alpha \in [0,1]$, which implies that $x_\alpha^- - \int_a^t f_\alpha^-(s) ds$ is non-decreasing with respect to $\alpha$ for $t \in \left[a, a + \frac{c_1}{M_1}\right]$. Similarly, since $(x_\alpha^+)' \leq c_2$ for all $\alpha \in [0,1]$ we have have

$$-\int_a^t \frac{\partial f_\alpha^+(s)}{\partial \alpha} ds \leq (t-a)M_2 \leq |c_2| \leq -(x_\alpha^+)'$$

and for any $t \in \left[a, a + \frac{|c_2|}{M_2}\right]$ and for all $\alpha \in [0,1]$, i.e. $x_\alpha^+ - \int_a^t f_\alpha^+(s)ds$ is non-increasing with respect to $\alpha$.

By the above reasoning it follows that $x \ominus \int_a^t f(s)ds$ exists, for all $t \in [a, h]$, where

$$h = \min\left\{\frac{c_1}{M_1}, \frac{|c_2|}{M_2}, \frac{len([x]_1)}{M}\right\} > 0.$$

If we are under the assumption b), then it follows that $len([f(s)]_1) = 0$, for all $s \in [a, b]$ and

$$\int_a^t len([f(s)]_1)ds = len([x]_1) = 0,$$

for all $t \in [a, b]$. The other two required inequalities can be obtained by similar reasoning as above, for all $t \in [a, a+h]$, where $h = \min\left\{\frac{c_1}{M_1}, \frac{|c_2|}{M_2}\right\} > 0$, which proves the lemma. ∎

The following result concerns the existence and uniqueness of solutions of a fuzzy differential equation under generalized differentiability.

**Theorem 9.10.** *(Bede-Gal [20], Bede-Gal [21])Let $R_0 = [t_0, t_0+p] \times \overline{B}(x_0, q)$, $p, q > 0$, $x_0 \in \mathbb{R}_\mathcal{F}$ and $f : R_0 \to \mathbb{R}_\mathcal{F}$ be continuous such that the following assumptions hold:*

*(i) There exists a constant $L > 0$ such that*

$$D(f(t,x), f(t,y)) \leq L \cdot D(x,y), \forall (t,x), (t,y) \in R_0.$$

*(ii) Let $[f(t,x)]_\alpha = [f_\alpha^-(t,x), f_\alpha^+(t,x)]$ be the level set representation of $f$, then $f_\alpha^-, f_\alpha^+ : R_0 \to \mathbb{R}$ have bounded partial derivatives with respect to $\alpha \in [0,1]$, the bounds being independent of $(t,x) \in R_0$ and $\alpha \in [0,1]$.*

*(iii) The functions $x_0^-$ and $x_0^+$ are differentiable (as functions of $\alpha$), and there exist $c_1 > 0$ with $(x_0)_\alpha^- \geq c_1$, and there exists $c_2 < 0$ with $(x_0)_\alpha^+ \leq c_2$, for all $\alpha \in [0,1]$ and we have the possibilities*

*a) $x_1^- < x_1^+$*

*or*

*b) if $x_1^- = x_1^+$ then the core $[f(t,x)]_1$ consists in exactly one element for any $(t,x) \in R_0$, whenever $[x]_1$ consist in exactly one element.*

*Then the fuzzy initial value problem*

$$x'(t) = f(t,x), x(t_0) = x_0,$$

*has exactly two solutions defined in an interval $[t_0, t_0 + k]$ for some $k > 0$.*

**Proof.** We use Lemma 9.2 together with the previous Lemma 9.9. The existence result in Theorem 9.3 ensures the existence of one solution which is Hukuhara differentiable. Here we will prove the existence of the other solution.

First, let us observe that assumptions (ii) and (iii), by Lemma 9.9 ensure the existence of the H-difference

$$x_0 \ominus \left( -\int_{t_0}^{t} F(t, x(t))dt \right)$$

for $t \in [t_0, t_0 + c]$ for some $0 < c \leq p$. Now we consider $R_1 = [t_0, t_0 + c] \times \overline{B}(x_0, q)$, $K_0 = C([t_0, t_0+c], \mathbb{R}_{\mathcal{F}})$ and the operator $Q : K_0 \to K_0$ ($C([a, b], \mathbb{R}_{\mathcal{F}})$ being the space of continuous functions $x : [a, b] \to \mathbb{R}_{\mathcal{F}}$) defined as follows:

$$Q(x_0)(t) = x_0$$

$$Q(x)(t) = x_0 \ominus \left( -\int_{t_0}^{t} F(t, x(t))dt \right).$$

We observe that $Q$ is well defined on $[t_0, t_0 + c]$ by the choice of $c$. By Lemma 9.2, and the Lipschitz condition (i) of the present Theorem, $f$ is bounded and we have

$$D(Q(x)(t), x_0) \leq \int_{t_0}^{t} D(f(t, x(t)), 0)dt \leq M(t - t_0),$$

where $M = \sup_{(t,x) \in R_1} D(f(t, x), 0)$ is provided by Lemma 9.2. Let $d = \min\left\{c, \frac{q}{M}\right\}$ and

$$K_1 = C([t_0, t_0 + d], \overline{B}(x_0, q)).$$

Then for the restriction $Q : K_1 \to C([t_0, t_0 + d], \mathbb{R}_{\mathcal{F}})$, we have

$$D(Q(x)(t), x_0) \leq q,$$

i.e., $x \in K_1$ gives $Q(x) \in K_1$, and $K_1$ is a complete metric space considered with the uniform distance, as a closed subspace of a complete metric space. Now we will show that $Q$ is a contraction. Indeed,

$$D(Q(x)(t), Q(y)(t)) \leq \int_{t_0}^{t} D(f(t, x(t)), f(t, y(t)))dt$$

$$\leq 2L(t - t_0)D(x, y).$$

Now choosing $k < \min\{d, \frac{1}{2L}\}$, $Q$ becomes a contractions. Banach's fixed point theorem implies the existence of a fixed point of $Q$. To conclude, we observe that the fixed point of $Q$ is strongly generalized differentiable as in case (ii) of Definition 8.31 and it solves the fuzzy initial value problem

$$x'(t) = f(t, x), x(t_0) = x_0,$$

for any $t \in [t_0, t_0 + k]$. The local uniqueness of this solution on $[t_0, t_0 + k]$ follows by the uniqueness of the fixed point for $Q$.    ∎

## 9.3.2    *Characterization Results*

We will extend in this section the Characterization result of Theorem 9.4 to the case of generalized differentiability.

**Theorem 9.11.** *(Bede [15], Bede-Gal [20]) Let $R_0 = [t_0, t_0 + p] \times \overline{B}(x_0, q)$, $p > 0$, $x_0 \in \mathbb{R}_{\mathcal{F}}$ and $f : R_0 \to \mathbb{R}_{\mathcal{F}}$ be such that*

$$[f(t,x)]_\alpha = [f_\alpha^-(t, x_\alpha^-, x_\alpha^+), f_\alpha^+(t, x_\alpha^-, x_\alpha^+)], \forall \alpha \in [0,1]$$

*and the following assumptions hold:*

*(i) $f_\alpha^\pm(t, x_\alpha^-, x_\alpha^+)$ are equicontinuous, uniformly Lipschitz in their second and third arguments i.e., there exist a constant $L > 0$ such that*

$$|f_\alpha^\pm(t, x_\alpha^-, x_\alpha^+) - f_\alpha^\pm(t, y_\alpha^-, y_\alpha^+)| \leq L(|x_\alpha^- - y_\alpha^-| + |x_\alpha^+ - y_\alpha^+|),$$

$\forall (t,x), (t,y) \in R_0, \alpha \in [0,1]$.

*(ii) $f_\alpha^-$, $f_\alpha^+ : R_0 \to \mathbb{R}$ have bounded partial derivatives with respect to $\alpha \in [0,1]$, the bounds being independent of $(t,x) \in R_0$ and $\alpha \in [0,1]$.*

*(iii) The functions $x_0^-$ and $x_0^+$ are differentiable, existing $c_1 > 0$ with $(x_0)_\alpha^- \geq c_1$, and $c_2 < 0$ with $(x_0)_\alpha^+ \leq c_2$, for all $\alpha \in [0,1]$, and we have the following possibilities*

*a) $(x_0)_1^- < (x_0)_1^+$*

*or*

*b) if $(x_0)_1^- = (x_0)_1^+$ then the core $[(t, x, u)]_1$ consists in exactly one element for any $(t,x) \in R_0$, whenever $[x]_1$ and $[u]_1$ consist in exactly one element.*

*Then the fuzzy initial value problem*

$$x'(t) = f(t, x), x(t_0) = x_0, \tag{9.1}$$

*is equivalent on some interval $[t_0, t_0 + k]$ with the union of the following two ODEs:*

$$\begin{cases} (x_\alpha^-)'(t) = f_\alpha^-(t, x_\alpha^-(t), x_\alpha^+(t)) \\ (x_\alpha^+)'(t) = f_\alpha^+(t, x_\alpha^-(t), x_\alpha^+(t)) & , \alpha \in [0,1] \\ x_\alpha^-(t_0) = (x_0)_\alpha^-, x_\alpha^+(t_0) = (x_0)_\alpha^+ \end{cases} \tag{9.2}$$

$$\begin{cases} (x_\alpha^-)'(t) = f_\alpha^+(t, x_\alpha^-(t), x_\alpha^+(t)) \\ (x_\alpha^+)'(t) = f_\alpha^-(t, x_\alpha^-(t), x_\alpha^+(t)) & , \alpha \in [0,1]. \\ x_\alpha^-(t_0) = (x_0)_\alpha^-, x_\alpha^+(t_0) = (x_0)_\alpha^+ \end{cases} \tag{9.3}$$

**Proof.** Condition (i) of the theorem ensures the existence of a unique solution for each of the systems of equations (9.2), (9.3) for any fixed $\alpha \in [0,1]$. Let us denote these solutions by $(x_\alpha^-)^i, (x_\alpha^+)^i$ and $(x_\alpha^-)^{ii}, (x_\alpha^+)^{ii}$ respectively. The equicontinuity of $f_\alpha^\pm$ ensures the continuity of $f$ as a fuzzy valued function, while the Lipschitz condition in (i) is sufficient for the corresponding

Lipschitz property of $f$. All these conditions together ensure the existence and uniqueness of two solutions $x^i$ and $x^{ii}$ for the fuzzy initial value problem (9.1). Let $[x^i]_\alpha = [(x^i)^-_\alpha, (x^i)^+_\alpha]$ and $[x^{ii}]_\alpha = [(x^{ii})^-_\alpha, (x^{ii})^+_\alpha]$, $\alpha \in [0,1]$. Then, since $x^i$ Hukuhara differentiable then

$$\left[ (x^i)' \right]_\alpha = [\left( (x^i)^-_\alpha \right)', \left( (x^i)^+_\alpha \right)']$$

and $(x^i)^-_\alpha, (x^i)^+_\alpha$ is a solution of the system (9.2). Since $(x^-_\alpha)^i, (x^+_\alpha)^i$ is the unique solution of the system we obtain

$$\left( (x^i)^-_\alpha, (x^i)^+_\alpha \right) = \left( (x^-_\alpha)^i, (x^+_\alpha)^i \right), \forall \alpha \in [0,1].$$

Similar reasoning for the (ii)-differentiable solution implies

$$\left[ (x^{ii})' \right]_\alpha = [\left( (x^{ii})^+_\alpha \right)', \left( (x^{ii})^-_\alpha \right)'],$$

so $(x^{ii})^+_\alpha, (x^{ii})^-_\alpha$ is a solution of the second system (9.3) considered in the Theorem, so

$$\left( (x^{ii})^-_\alpha, (x^{ii})^+_\alpha \right) = \left( (x^-_\alpha)^{ii}, (x^+_\alpha)^{ii} \right), \forall \alpha \in [0,1].$$

As a conclusion the solutions of the initial value problem (9.1) are exactly those of the given systems (9.2) and (9.3). ∎

In the followings we formulate a particularization of the existence, uniqueness and characterization Theorem 9.11 for fuzzy initial value problems with triangular data. We denote by $\mathbb{R}_\mathcal{T}$ the space of triangular fuzzy numbers.

**Theorem 9.12.** *Let $R_0 = [t_0, t_0 + p] \times (\overline{B}(x_0, p) \cap \mathbb{R}_\mathcal{T}), p > 0, x_0 \in \mathbb{R}_\mathcal{T}$ and $f : R_0 \to \mathbb{R}_\mathcal{T}$ be such that*

$$f(t,x) = (f^-(t, x^-, x^1, x^+), f^1(t, x^-, x^1, x^+), f^+(t, x^-, x^1, x^+))$$

*and the following assumptions hold:*

*(i) $f^-, f^1, f^+$ are continuous, Lipschitz in their second through last arguments.*

*(ii) $x_0 = (x^-_0, x^1_0, x^+_0)$ is a nontrivial triangular number such that $x^-_0 < x^1_0 < x^+_0$. Then*

*(a) The fuzzy initial value problem*

$$x'(t) = f(t,x), x(t_0) = x_0, \tag{9.4}$$

*has exactly two triangular valued solutions on some interval $[t_0, t_0 + k]$.*

*(b) Problem (9.4) is equivalent to the union of the following two ODEs:*

$$\begin{cases} (x^-)' = f^-(t, x^-, x^1, x^+) \\ (x^1)' = f^1(t, x^-, x^1, x^+) \\ (x^+)' = f^+(t, x^-, x^1, x^+) \\ x^-(t_0) = x_0^-, x^1(t_0) = x_0^1, x^+(t_0) = x_0^+ \end{cases} , \qquad (9.5)$$

$$\begin{cases} (x^-)' = f^+(t, x^-, x^1, x^+) \\ (x^1)' = f^1(t, x^-, x^1, x^+) \\ (x^+)' = f^-(t, x^-, x^1, x^+) \\ x^-(t_0) = x_0^-, x^1(t_0) = x_0^1, x^+(t_0) = x_0^+ \end{cases} , \qquad (9.6)$$

**Proof.** It is easy to check that since $f$ is triangular valued, the solutions of (9.4) are triangular valued. Further the conditions (i) and (ii) ensure that the problem (9.4) has two unique solutions locally. It is easy to check that the conditions in Theorem 9.11 are fulfilled and the problems (9.2) and (9.3), in the case of triangular-valued functions are equivalent with (9.5) and (9.6) respectively, and the proof is complete.    ∎

### 9.3.3   Examples of Fuzzy Differential Equations under Strongly Generalized Differentiability

In what follows we solve several fuzzy differential equations under generalized differentiability.

**Example 9.13.** *Let us begin with the homogeneous decay equation*

$$x' = -x, \ x(0) = (-1, 0, 1).$$

*Surely since (i)-differentiability is in fact Hukuhara differentiability we obtain the solution $x(t) = e^t(-1, 0, 1)$. Under (ii)-differentiability condition we get the solution $x(t) = e^{-t}(-1, 0, 1)$. The behaviors of the two solutions differ substantially, as shown in Fig. 9.3.*

**Example 9.14.** *By adding a forcing term we obtain the following equation (see Tenali-Lakshmikantham-Devi [145] and Bede-Tenali-Lakshmikantham [26])*

$$x' = -x + e^{-t}(-1, 0, 1).$$

*Let us keep the initial condition the same as in the previous problem, i.e. $x_0 = (-1, 0, 1)$ under strongly generalized differentiability cases (i) or (ii).*
   *For the case of Hukuhara or (i)-differentiability we get*

$$\begin{cases} (x_0^-)' = -x_0^+ - e^{-t} \\ x_1' = -x_1 \\ (x_0^+)' = -x_0^- + e^{-t} \end{cases}$$

*and it has the solution $x(t) = \left( \frac{1}{2}e^{-t} - \frac{3}{2}e^t, 0, \frac{3}{2}e^t - \frac{1}{2}e^{-t} \right)$.*

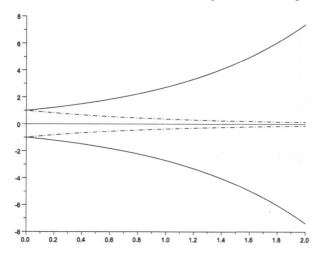

**Fig. 9.3** Two solutions of a fuzzy differential equation (solid line is (i) differentiable solution, dash-dot line is the (ii) differentiable solution

*Under (ii)-differentiability, the fuzzy differential equation becomes*

$$\begin{cases} (x_0^-)' = -x_0^- + e^{-t} \\ x_1' = -x_1 \\ (x_0^+)' = -x_0^+ - e^{-t} \end{cases}$$

*and we get $x(t) = e^{-t}(1-t)(-1,0,1)$. In Fig.9.4 the (i)-differentiable and (ii)-differentiable solutions are presented.*

**Example 9.15.** *This example was proposed in Bede-Gal [21]. In Barros-Bassanezi [13] and Nieto-Rodriguez-Lopez [117] a logistic fuzzy model is analyzed in detail and the findings are that it is an adequate model for population dynamics under uncertainty. We use the strongly generalized differentiability and the results on characterization previously discussed to analyze a fuzzy logistic equation*

$$x' = ax(K \ominus_{gH} x), \; x(0) = x_0$$

*where $aKx$ is the growth rate, $ax^2$ is the inhibition term and $K$ is the environment capacity. The generalized Hukuhara difference $K \ominus_{gH} x$ is used here. The advantage of this difference over the Hukuhara difference in this example is that the generalized Hukuhara difference exists in situations when the usual Hukuhara difference fails to exist. We consider $a, K \in \mathbb{R}_{\mathcal{T}}$ symmetric triangular numbers and $x_0, K \in \mathbb{R}$. We use in our example $x_0 = 500$, $a = (0.98, 1, 1.02) \cdot 10^{-3}$, $K = (9, 10, 11) \cdot 10^3$. Let us remark that the*

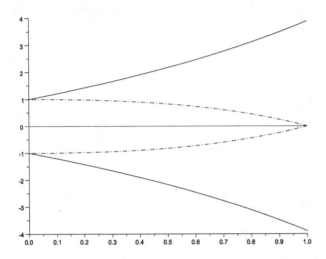

**Fig. 9.4** Two solutions of a fuzzy differential equation (solid line is (i) differentiable solution, dash-dot line is the (ii) differentiable solution

*function $f(t, x) = ax(K \ominus_g x)$ is not a triangular-number-valued function even if the inputs are triangular numbers, since the product of two triangular fuzzy numbers is not generally triangular. In this case we have for any $\alpha \in [0,1]$,*

$$F_\alpha^- (x^-, x^1, x^+) = a_\alpha^- \min_{\alpha \in [0,1]} \left\{ x_\alpha^- \left( K_\alpha^- - x_\alpha^- \right), x_\alpha^+ \left( K_\alpha^+ - x_\alpha^+ \right) \right\}$$

$$F^1 (x^-, x^1, x^+) = a^1 x^1 \left( K^1 - x^1 \right)$$

$$F_\alpha^+ (x^-, x^1, x^+) = a_\alpha^+ \max_{\alpha \in [0,1]} \left\{ x_\alpha^- \left( K_\alpha^- - x_\alpha^- \right), x_\alpha^+ \left( K_\alpha^+ - x_\alpha^+ \right) \right\}.$$

*These are piecewise quadratic functions of $\alpha$ and the more general characterization result 9.11 is necessarily used. Locally, the FIVP is equivalent to the problem*

$$\begin{cases} (x_\alpha^-)' = F_\alpha^- (t, x^-, x^1, x^+) \\ (x_\alpha^+)' = F_\alpha^+ (t, x^-, x^1, x^+) \\ x^-(t_0) = x_0^-, x^1(t_0) = x_0^1, x^+(t_0) = x_0^+ \end{cases}. \tag{9.7}$$

*It is easy to see that the only solution we have locally on some interval $[0, p]$ is the solution according to case (i) of differentiability, since $x_0 \in \mathbb{R}$. Let $p = \inf\{t > 0 : x_\alpha^-(t) (K_\alpha^- - x_\alpha^-(t)) = x_\alpha^+(t)(K_\alpha^+ - x_\alpha^+(t))\}$. We have numerically obtained $p \approx 0.407$. We observe that if we use now $p$ and $x_0 = x(p)$ as initial values for a new initial value problem*

$$x' = ax(K \ominus_{gH} x), x(p) = x_0$$

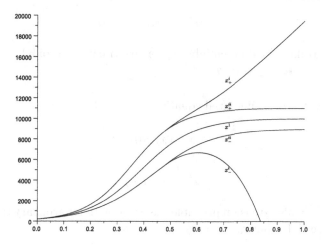

**Fig. 9.5** Two solutions of a fuzzy differential equation

*two solutions will emerge according to Theorem 9.10 characterized by the
systems (9.7) and (9.8),*

$$\begin{cases} (x_\alpha^-)' = F_\alpha^-(t, x^-, x^1, x^+) \\ (x_\alpha^+)' = F_\alpha^+(t, x^-, x^1, x^+) \\ x^-(t_0) = x_0^-, x^1(t_0) = x_0^1, x^+(t_0) = x_0^+ \end{cases} \qquad (9.8)$$

*One is differentiable as in (i) and another differentiable as in (ii) case of Definition 8.31. Let us remark also, that we can paste together the solutions and
we obtain two solutions on the $[0, 1]$ interval. We can see that the most realistic model appears to be the one which starts with case (i) of differentiability
on $[0, p]$ and continues with case (ii) of differentiability on $[p, 1]$. The solution
which is (i) differentiable is the Hukuhara solution on $[0, 1]$ and it is unrealistic from practical point of view since the population may increase according
to that solution much more than the environment's capacity. In Figure 9.5
these solutions are shown, where $x^i$ represents the lower and upper endpoints
of the 0-level set of the solution according to case (i) of differentiability, $x^{ii}$
is the 0 level set of the solution as in case (ii) of differentiability. $x^1$ is the
common 1-level set of the two solutions.*

*This example also allows us to underline the practical superiority of the
strongly generalized differentiability with respect to the Hukuhara derivative
case. Several other practical examples were proposed in the literature.*

## 9.4   Problems

1. Solve the fuzzy differential equations

$$x' = x, x(0) = x_0$$

$$x' = -x, x(0) = x_0$$

under Hukuhara differentiability, for arbitrary trapezoidal fuzzy initial value $x_0 \in \mathbb{R}_{\mathcal{F}}$ $x_0 = (a, b, c, d)$.

2. Solve the fuzzy differential equations

$$x' = x, x(0) = x_0$$

$$x' = -x, x(0) = x_0$$

using Zadeh's extension principle-based method, for arbitrary trapezoidal fuzzy initial value $x_0 \in \mathbb{R}_{\mathcal{F}}$ $x_0 = (a, b, c, d)$.

3. Solve the fuzzy differential equations

$$x' = x, x(0) = x_0$$

$$x' = -x, x(0) = x_0$$

under generalized differentiability, for arbitrary trapezoidal fuzzy initial value $x_0 \in \mathbb{R}_{\mathcal{F}}$ $x_0 = (a, b, c, d)$.

4. Using Zadeh extension principle-based approach solve the fuzzy differential equation

$$x' = a \cdot x + b, x(0) = x_0$$

under strongly generalized differentiability with $a \in \mathbb{R}$, $b, x_0 \in \mathbb{R}_{\mathcal{F}}$, $b = (b^-, b^1, b^+)$, $x_0 = (x_0^-, x_0^1, x_0^+)$ triangular fuzzy numbers.

5. Solve the fuzzy differential equation

$$x' = a \cdot x + b, x(0) = x_0$$

under strongly generalized differentiability with $a \in \mathbb{R}$, $b, x_0 \in \mathbb{R}_{\mathcal{F}}$, $b = (b^-, b^1, b^+)$, $x_0 = (x_0^-, x_0^1, x_0^+)$ triangular fuzzy numbers.
Try to consider a generalization to a triangular number $a = (a^-, a^1, a^+)$, with a restriction to the case when $a^- > 0$ and $x_0^- > 0$.

6. State and prove a more general version of the Characterization result in Theorem 9.11, for a system of fuzzy differential equations and respectively for a second order fuzzy differential equation.

7. Consider a first order linear differential equation

$$y' = a(x) \cdot y + b(x), y(x_0) = y_0$$

with $a(x)$ real valued function, and $b : \mathbb{R} \to \mathbb{R}_{\mathcal{F}}$. Prove that

$$y_1(x) = e^{\int_{x_0}^x a(t)dt} \left( y_0 + \int_{x_0}^x b(t) \cdot e^{-\int_{x_0}^t a(u)du} dt \right) \qquad (9.9)$$

and

$$y_2(x) = e^{\int_{x_0}^x a(t)dt} \left( y_0 \ominus \int_{x_0}^x (-b(t)) \cdot e^{-\int_{x_0}^t a(u)du} dt \right), \qquad (9.10)$$

provided that the H-difference $y_0 \ominus \int_{x_0}^x (-b(t)) \cdot e^{-\int_{x_0}^t a(u)du} dt$ exists, are solutions of the fuzzy differential equation $y' = a(x) \cdot y + b(x)$.

8. Describe Euler's method to calculate the solution of a fuzzy differential equation using the characterization theorem presented in this chapter.

# 10

# Extensions of Fuzzy Set Theory

In fuzzy set theory the membership function of a fuzzy set is a classical function $A : X \to [0,1]$. In some applications the shape of the membership function is itself uncertain. This problem appears mainly because of the subjectivity of expert knowledge and imprecision of our models. In these situations we can use a higher order extension of fuzzy set theory.

## 10.1  Lattice Valued Fuzzy Sets (L-Fuzzy Sets)

**Definition 10.1.** *Let $(L, \leq)$, be a complete lattice (i.e., any non-empty subset of $L$ has an infimum and a supremum). A negation on a complete lattice is an operator $N : L \to L$ such that $N$ is involutive, i.e.*

$$N(N(x)) = x, \forall x \in L,$$

*and non-increasing, i.e., if $x \leq y$ then $N(x) \geq N(y)$, $x, y \in L$.*

In the followings we assume that $L$ is a complete lattice with a negation.

**Definition 10.2.** *(Goguen [73]) An L-fuzzy set on the universe $X$ is a mapping $A : X \to L$. The class of L-fuzzy sets is denoted by $\mathcal{F}_L(X)$. The operations on L-fuzzy sets are point-wise provided by the inf and sup that come from the lattice structure. The basic connectives of L-fussy sets are*

$$A \wedge B(x) = \inf(A(x), B(x)),$$

$$A \vee B(x) = \sup(A(x), B(x)),$$

B. Bede: *Mathematics of Fuzzy Sets and Fuzzy Logic*, STUDFUZZ 295, pp. 193–199.
DOI: 10.1007/978-3-642-35221-8_10      © Springer-Verlag Berlin Heidelberg 2013

$$N(A)(x) = N(A(x)),$$

*for any $x \in X$.*

In general, for L-fuzzy sets distributivity does not hold. In order to be able to use this property we need to have a distributive lattice i.e., such that the sup is distributive w.r.t. inf and vice versa

$$a \vee (b \wedge c) = (a \vee b) \wedge (a \vee c)$$

and

$$a \wedge (b \vee c) = (a \wedge b) \vee (a \wedge c).$$

Also, in some problems complete distributivity could be required i.e.,

$$a \vee \left( \bigwedge_{i \in I} b_i \right) = \bigwedge_{i \in I} a \vee b_i \text{ and } a \wedge \left( \bigvee_{i \in I} b_i \right) = \bigvee_{i \in I} a \wedge b_i.$$

If $L$ is a distributive (completely distributive) lattice, then the L-fuzzy set operations become distributive (completely distributive).

**Example 10.3.** *$([0,1], \leq)$ is a completely distributive lattice. Then classical fuzzy sets can be seen as L-fuzzy sets with $L = [0,1]$. We observe that $[0,1]$ is complete and completely distributive lattice. We observe that the L-fuzzy sets with $L = [0,1]$ coincide with the fuzzy sets $\mathcal{F}(X)$.*

**Example 10.4.** *$(\mathcal{F}(X), \leq)$ the family of fuzzy sets over the same universe of discourse $X$ with the inclusion. Then $\mathcal{F}(X)$ is a lattice. The infimum is the fuzzy intersection $\wedge$, while the supremum will be the fuzzy union $\vee$. We can now use $L = \mathcal{F}(X)$ as a lattice to define L-fuzzy sets. Surely iteration on this procedure is possible.*

## 10.2    Intuitionistic Fuzzy Sets

**Definition 10.5.** *(Atanassov [6]) Let $X$ be a non-empty set. An intuitionistic fuzzy set is a pair of mappings $A = (\mu_A, \nu_A) : X \to [0,1]$ that fulfill the condition*

$$\mu_A + \nu_A \leq 1, \forall x \in X.$$

*$\mu_A$ represents the degree of membership while $\nu_A$ is the degree of non-membership.*

**Remark 10.6.** *A fuzzy set $\mu_A$ can be interpreted as the intuitionistic fuzzy set $A = (\mu_A, 1 - \mu_A)$. The reciprocal is not necessarily true.*

The basic operations between intuitionistic fuzzy sets are defined as follows:

$$A \cap B = (\min(\mu_A, \mu_B), \max(\nu_A, \nu_B))$$

$$A \cup B = (\max(\mu_A, \mu_B), \min(\nu_A, \nu_B))$$

$$N(A) = (\nu_A, \mu_A).$$

We denote the collection of all Intuitionistic fuzzy sets by $IFS(X)$. Intuitionistic fuzzy sets were studied by several authors (e.g. Szmidt-Kacprzyk [143], Ban [7]) and they were used successfully in applications (Sussner et. al. [140] etc.)

**Proposition 10.7.** *The set $IFS(X)$ with the operations of union intersection and complement defined as above is a distributive lattice. Moreover it is complete and completely distributive.*

**Proof.** The proof is immediate.                                                ∎

**Remark 10.8.** *Intuitionistic fuzzy sets can be seen as L-fuzzy sets by considering the lattice $L \subseteq [0,1]^2$,*

$$L = \{(x,y) \in [0,1]^2 | x + y \leq 1\}$$

*where the inequality relation generating the Lattice structure is defined by*

$$(x,y) \leq_L (z,t) \Leftrightarrow [x \leq z \text{ and } y \geq t].$$

*Moreover, $L$ in this case is a complete, completely distributive lattice.*

## 10.3   Interval Type II Fuzzy Sets

Let $\mathbb{I}[0,1]$ denote the set of closed sub-intervals of $[0,1]$.

**Definition 10.9.** *(see e.g. Mendel, [110]) An interval valued fuzzy set is a mapping $A : X \rightarrow \mathbb{I}[0,1]$. Let us denote*

$$A(x) = [\underline{A}(x), \bar{A}(x)].$$

*Then $\underline{A}(x), \bar{A}(x)$ can be interpreted as the lower and upper bounds of the membership grade.*

Operations with interval valued fuzzy sets are defined by:

$$A \cup B(x) = [\sup\{\underline{A}(x), \underline{B}(x)\}, \sup\{\bar{A}(x), \bar{B}(x)\}]$$

$$A \cap B(x) = [\inf\{\underline{A}(x), \underline{B}(x)\}, \inf\{\bar{A}(x), \bar{B}(x)\}]$$

$$N A(x) = [1 - \bar{A}(x), 1 - \underline{A}(x)].$$

If $A = (\mu_A, \nu_A)$ is an intuitionistic fuzzy set then we define an interval valued fuzzy set, $\underline{A}(x) = \mu_A(x)$ and $\bar{A}(x) = 1 - \nu_A(x)$. Then we have a correspondence between interval valued fuzzy sets and intuitionistic fuzzy sets because we can reciprocally consider $\mu_A(x) = \underline{A}(x)$ and $\nu_A(x) = 1 - \bar{A}(x)$.

This correspondence is well defined since

$$\underline{A}(x) \leq \bar{A}(x)$$

$$\Leftrightarrow 1 - \bar{A}(x) \leq 1 - \underline{A}(x)$$

$$\Leftrightarrow \nu_A(x) \leq 1 - \mu_A(x)$$

$$\Leftrightarrow \mu_A(x) + \nu_A(x) \leq 1.$$

**Proposition 10.10.** *The set of interval valued fuzzy sets with the operations of union intersection and complement defined as above is a distributive lattice. Moreover it is complete and completely distributive.*

**Proof.** The proof is a direct consequence of the above discussion with Proposition 10.7. ∎

**Remark 10.11.** *Interval valued fuzzy sets can be seen as L-fuzzy sets, by considering the complete distributive lattice $L = \mathbb{I}[0,1]$ with the inequality defined as*

$$[a, b] \leq [c, d] \Leftrightarrow a \leq c \text{ and } b \leq d.$$

## 10.4   Fuzzy Sets of Type 2

Type $m$ fuzzy sets can be defined recursively as follows.

**Definition 10.12.** *(see e.g. Mendel, [110]) Type 1 fuzzy sets are regular fuzzy sets*

$$\mathcal{F}_1(X) = \{A : X \rightarrow [0,1]\}.$$

*Type $m$ fuzzy sets, $m > 1$ are L-fuzzy sets whose membership degrees are type $m - 1$ fuzzy sets*

$$\mathcal{F}_m(X) = \{A : X \rightarrow \mathcal{F}_{m-1}([0,1])\}, m \geq 2.$$

Type 2 fuzzy sets are the most important cases of type $m$ fuzzy sets. They are characterized by membership function $\mu_A : X \rightarrow \mathcal{F}[0,1]$.

$$\mu_A(x)(u) \in [0,1],$$

$\forall x \in X$ and $u \in [0,1]$.

For a given $x \in X$ we denote by $J_x$ the support of $\mu_A(x)$, i.e.,

$$J_x = \{u : |\mu(x)(u) \neq 0\}.$$

The notations used for fuzzy sets of type 2 are as follows

$$\tilde{A} = \{(x, u, \mu_{\tilde{A}}(x, u)) | x \in X, u \in J_X\},$$

$$\tilde{A} = \int_{x \in X,\, u \in J_x} \int \mu_{\tilde{A}}(x, u)/(x, u), J_x \subseteq [0, 1],$$

or

$$\tilde{A}(x) = \int_{u \in J_x} \mu_{\tilde{A}}(x)(u)/u, J_x \subseteq [0, 1],$$

where the integral sign denotes union over all admissible values.

There is a strong connection between fuzzy sets of type 2 and interval valued fuzzy sets. Namely, if we consider a fuzzy set of type 2 $\tilde{A}$ then its footprint of uncertainty is defined as

$$FOU(\tilde{A}) = \bigcup_{x \in X} J_x.$$

We can determine two type 1 fuzzy sets

$$\underline{\mu}_{\tilde{A}}(x) = \underline{J_x} = \inf J_x, \bar{\mu}_{\tilde{A}}(x) = \bar{J_x} = \sup J_x.$$

Now we define $I_x = [\underline{J_x}, \bar{J_x}]$, and in this way we have obtained an interval valued fuzzy set $B : X \to \mathbb{I}[0, 1], B(x) = I_x$.

As a conclusion from a type 2 fuzzy set we can obtain an interval valued fuzzy set. The reciprocal is not true, i.e. from an interval valued fuzzy set we cannot obtain the type 2 fuzzy set that was initially considered. This shows that type 2 fuzzy sets are more general than interval valued fuzzy sets or intuitionistic fuzzy sets.

The operations between interval type 2 fuzzy sets can be directly defined as follows.

$$\tilde{A} \cup \tilde{B}(x)(u) = \mu_{\tilde{A}}(x, u) \vee \mu_{\tilde{B}}(x, u)$$

$$\tilde{A} \cap \tilde{B}(x)(u) = \mu_{\tilde{A}}(x, u) \wedge \mu_{\tilde{B}}(x, u)$$

$$N(\tilde{A})(x)(u) = 1 - \mu_{\tilde{A}}(x, u).$$

These operations do not have some desirable properties and interpretation. For example they do not satisfy Zadeh's extension principle. Different operations are introduced and studied in the literature, based on Zadeh's extension principle as follows:

$$\tilde{A} \vee \tilde{B}(x)(u) = \sup\{\min(\mu_{\tilde{A}}(x, v), \mu_{\tilde{B}}(x, w)) | \max(v, w) = u\}$$

$$\tilde{A} \wedge \tilde{B}(x)(u) = \sup\{\min(\mu_{\tilde{A}}(x, v), \mu_{\tilde{B}}(x, w)) | \min(v, w) = u\}$$

$$N(\tilde{A})(x)(u) = 1 - \mu_{\tilde{A}}(x, u).$$

## 10.5  Problems

1. Prove that the operations defined on intuitionistic fuzzy sets verify De Morgan laws:
$$N(A \vee B) = N(A) \wedge N(B)$$

2. Consider the following intuitionistic fuzzy set
$$A(x) = (\mu_A(x), \nu_A(x)),$$

where $\mu_A(x) = (7, 8, 9)$ and

$$\nu_A(x) = \begin{cases} 1 & \text{if} & x < 6 \\ \frac{8-x}{2} & \text{if} & 6 \le x < 8 \\ \frac{x-8}{2} & \text{if} & 8 \le x \le 10 \\ 0 & \text{if} & 10 < x \end{cases}.$$

Verify that the membership functions considered are defining indeed an intuitionistic fuzzy set.

3. Prove that $L = \mathbb{I}[0, 1]$ with the inequality defined as $[a, b] \le [c, d] \iff a \le c$ and $b \le d$ is a lattice. Find the supremum and infimum of two intervals $[0.2, 0.5]$, $[0.3, 0.6]$.

4. Consider the following interval valued fuzzy sets $A(x) = [\underline{A}(x), \bar{A}(x)]$ and $B(x) = [\underline{B}(x), \bar{B}(x)]$ where

$$\underline{A}(x) = \begin{cases} 0 & \text{if} & x < 2 \\ x - 2 & \text{if} & 2 \le x < 3 \\ 4 - x & \text{if} & 3 \le x \le 4 \\ 0 & \text{if} & 4 < x \end{cases},$$

$$\bar{A}(x) = \begin{cases} 0 & \text{if} & x < 1 \\ \frac{x-1}{2} & \text{if} & 1 \le x < 3 \\ \frac{5-x}{2} & \text{if} & 3 \le x \le 5 \\ 0 & \text{if} & 5 < x \end{cases}.$$

$$\underline{B}(x) = \begin{cases} 0 & \text{if} & x < 4 \\ x - 4 & \text{if} & 4 \le x < 5 \\ 6 - x & \text{if} & 5 \le x \le 6 \\ 0 & \text{if} & 6 < x \end{cases},$$

$$\bar{B}(x) = \begin{cases} 0 & \text{if} & x < 3 \\ \frac{x-3}{2} & \text{if} & 3 \leq x < 5 \\ \frac{7-x}{2} & \text{if} & 5 \leq x \leq 7 \\ 0 & \text{if} & 7 < x \end{cases}.$$

Verify that the given membership degrees define an interval valued fuzzy set by sketching the membership functions. Calculate $A \wedge B$ and $A \vee B$ and graph them.

5. Given interval valued fuzzy sets $A(x) = [\underline{A}(x), \bar{A}(x)]$ and $B(x) = [\underline{B}(x), \bar{B}(x)]$ with

$$\underline{A}(x) = \begin{cases} 0 & \text{if} & x < 2 \\ x - 2 & \text{if} & 2 \leq x < 3 \\ 4 - x & \text{if} & 3 \leq x \leq 4 \\ 0 & \text{if} & 4 < x \end{cases},$$

$$\bar{A}(x) = \begin{cases} 0 & \text{if} & x < 1 \\ \frac{x-1}{2} & \text{if} & 1 \leq x < 3 \\ \frac{x-t}{2} & \text{if} & 3 \leq x \leq 5 \\ 0 & \text{if} & 5 < x \end{cases}.$$

and

$$\underline{B}(x) = \begin{cases} 0 & \text{if} & x < 20 \\ \frac{x-20}{10} & \text{if} & 20 \leq x < 30 \\ \frac{40-x}{10} & \text{if} & 30 \leq x \leq 40 \\ 0 & \text{if} & 40 < x \end{cases},$$

$$\bar{B}(x) = \begin{cases} 0 & \text{if} & x < 10 \\ \frac{x-10}{20} & \text{if} & 10 \leq x < 30 \\ \frac{50-x}{20} & \text{if} & 30 \leq x \leq 50 \\ 0 & \text{if} & 50 < x \end{cases}.$$

we consider the fuzzy rule

$$\text{if } x \text{ is } A \text{ then } y \text{ is } B$$

with a Mamdani inference system. Given the crisp input $x = 2.5$ find the firing level (interval) for the given fuzzy rule. Sketch the graph of the output of the inference system.

6. Prove Proposition 10.7.

# 11

# Possibility Theory

## 11.1 Fuzzy Sets and Possibility Distributions

Let $X$ be a set, adopted as the universe of discourse in what follows. We ask the question whether a value for a variable is possible. The answer will be a fuzzy set when the problem we are considering has uncertainties that are not of statistical type.

**Definition 11.1.** *(Dubois-Prade [51]) A possibility distribution $\pi$ of a fuzzy variable $x$ on a set of possible situations $X$ is a fuzzy set $\pi : X \rightarrow [0,1]$, such that there exists an $x_0 \in X$ such that $\pi(x_0) = 1$.*

A possibility distribution aims to distinguish between what is expected and what is surprising, with the following convention $\pi(u) = 0$ means that $x = u$ is rejected as impossible, $\pi(u) = 1$ is totally possible, while $\pi(u) = 0.7$ means that it is possible with the degree 0.7. So, $\pi(u)$ means the degree of possibility of the assignment $x = u$, some values $u$ being more possible.

Now we can ask the question: If we consider a crisp set of values $A \subseteq X$ what is the possibility degree of $x \in A$. Also, a related question is the following: Does $x$ necessarily belong to $A$?

**Definition 11.2.** *(Dubois-Prade [51]) The possibility degree of $x$ to belong to $A$ is*

$$\Pi(A) = \sup_{x \in A} \pi(x).$$

*The necessity degree is*

$$N(A) = \inf_{x \in A} n(\pi(A)),$$

*where $n$ is the standard negation.*

B. Bede: *Mathematics of Fuzzy Sets and Fuzzy Logic*, STUDFUZZ 295, pp. 201–212.
DOI: 10.1007/978-3-642-35221-8_11     © Springer-Verlag Berlin Heidelberg 2013

The mathematical theory for possibility distributions is Fuzzy Measure Theory.

## 11.2   Fuzzy Measures

**Definition 11.3.** *(Sugeno-Murofushi [139]) Let $\mu$ be a function from a $\sigma-$ algebra $\mathcal{B} \subseteq \mathcal{P}(X)$ to $[0,1]$. $\mu$ is a fuzzy measure if*
  *(i) $\mu(\emptyset) = 0, \mu(X) = 1$*
  *(ii) For any $A, B \in \mathcal{B}$ we have $\mu(A) \leq \mu(B)$.*
  *(iii) If $\{A_i\}_{i=1,2,\dots} \subseteq \mathcal{B}$ is an ascending (respectively descending) monotone sequence $A_1 \subseteq A_2 \subseteq \dots \subseteq A_n \subseteq \dots$ (respectively $A_1 \supseteq A_2 \supseteq \dots \supseteq A_n \supseteq \dots$) then*

$$\lim_{n \to \infty} \mu(A_n) = \mu(\lim_{n \to \infty} A_n),$$

*where $\lim_{n \to \infty} A_n = \cup_{n=1}^{\infty} A_n$ (respectively $\lim_{n \to \infty} A_n = \cap_{n=1}^{\infty} A_n$)*

**Proposition 11.4.** *(Gal-Ban [70])For a fuzzy measure we have*
  *(i) $\mu(A \cup B) \geq \max(\mu(A), \mu(B))$*
  *(ii) $\mu(A \cap B) \leq \min(\mu(A), \mu(B))$*

**Proof.** Since

$$A \cap B \subseteq A \subseteq A \cup B$$
$$A \cap B \subseteq B \subseteq A \cup B$$

from the monotonicity of $g$ the statement easily follows.    ∎

**Example 11.5.** *Probability measures are fuzzy measures. The additional property that they have is that of $\sigma-$additivity*

$$P\left(\bigcup_{i=1}^{\infty} A_i\right) = \sum_{i=1}^{\infty} P(A_i),$$

*where $A_i \cap A_j = \emptyset$ when $i \neq j$.*

**Example 11.6.** *Dirac measure*

$$\mu(A) = \begin{cases} 1 & if \quad x_0 \in A \\ 0 & otherwise \end{cases},$$

*where $x_0 \in X$ is fixed, is a fuzzy measure.*

**Definition 11.7.** *(Gal-Ban [70]) A set function $g_\lambda : \mathcal{B} \to [0,1]$ that fulfills*
  *(i) $g_\lambda(X) = 1$*
  *(ii)*

$$g_\lambda(A \cup B) = g_\lambda(A) + g_\lambda(B) + \lambda \cdot g_\lambda(A) \cdot g_\lambda(B)$$

*with $A \cap B = \emptyset$, is called a Sugeno $\lambda-$measure.*

**Remark 11.8.** *If $\lambda = 0$ then a Sugeno $\lambda$−measure becomes a classical measure.*

**Lemma 11.9.** *If $B_i$ are disjoint sets and $\lambda \neq 0$ then*

$$g_\lambda \left( \bigcup_{i=1}^{\infty} B_i \right) = \frac{1}{\lambda} \left[ \prod_{i=1}^{\infty} (1 + \lambda g_\lambda(B_i)) - 1 \right].$$

**Proof.** The relation

$$g_\lambda \left( \bigcup_{i=1}^{n} B_i \right) = \frac{1}{\lambda} \left[ \prod_{i=1}^{n} (1 + \lambda g_\lambda(B_i)) - 1 \right],$$

with arbitrary $n \geq 1$ can be proved by induction. By taking the limit as $n \to \infty$ we obtain the relation in the statement of the Lemma. ∎

**Lemma 11.10.** *If $g_\lambda$ is a $\lambda$−fuzzy measure then*

$$g_\lambda(\bar{A}) = \frac{1 - g_\lambda(A)}{1 + \lambda g_\lambda(A)}.$$

**Proof.** The proof is left to the reader as an exercise. ∎

**Proposition 11.11.** *(Gal-Ban [70]) A Sugeno $\lambda$−measure is a fuzzy measure for $\lambda > -1$.*

**Proof.** (i) We have

$$g_\lambda(X \cup \emptyset) = g_\lambda(X) + g_\lambda(\emptyset) + \lambda g_\lambda(X) g_\lambda(\emptyset).$$

Since $g_\lambda(X) = 1$ we get $g_\lambda(\emptyset)(1 + \lambda) = 0$ and then $g_\lambda(\emptyset) = 0$, since $\lambda > -1$.

(ii) Let $A \subseteq B$. Then there exist $C = B \setminus A$ with $A \cup C = B$ and $A \cap C = \emptyset$. Then

$$g_\lambda(B) = g_\lambda(A \cup C) = g_\lambda(A) + g_\lambda(C) + \lambda g_\lambda(A) g_\lambda(C)$$
$$= g_\lambda(A) + (1 + \lambda g_\lambda(A)) g_\lambda(C) \geq g_\lambda(A).$$

(iii) Let $A_1 \subseteq A_2 \subseteq ... \subseteq A_n \subseteq ...$ be an ascending sequence. Let $B_1 = A_1, B_2 = A_2 \setminus B_1, ..., B_n = A_n \setminus B_{n-1}, ...$. Then $B_i \cap B_j = \emptyset, i \neq j$ and $\cup_{n=1}^{\infty} A_n = \cup_{n=1}^{\infty} B_n$.

Taking into account Lemma 11.9 we obtain

$$g_\lambda( \lim_{n \to \infty} A_n) = g_\lambda \left( \bigcup_{i=1}^{\infty} A_i \right) = g_\lambda \left( \bigcup_{i=1}^{\infty} B_i \right)$$
$$= \frac{1}{\lambda} \left[ \prod_{i=1}^{\infty} (1 + \lambda g_\lambda(B_i)) - 1 \right].$$

On the other hand since $A_n = \cup_{i=1}^n B_i$ we obtain

$$\lim_{n\to\infty} g_\lambda(A_n) = \lim_{n\to\infty} g_\lambda\left(\bigcup_{i=1}^n B_i\right)$$

$$= \lim_{n\to\infty} \frac{1}{\lambda}\left[\prod_{i=1}^n (1 + \lambda g_\lambda(B_i)) - 1\right]$$

Comparing the two relations above we get

$$g_\lambda(\lim_{n\to\infty} A_n) = \lim_{n\to\infty} g_\lambda(A_n),$$

i.e., the continuity by ascending sequences.

For the continuity by descending sequences we consider $A_1 \supseteq A_2 \supseteq \dots \supseteq A_n \supseteq \dots$. Then the complements $\bar{A}_1 \subseteq \bar{A}_2 \subseteq \dots \subseteq \bar{A}_n \subseteq \dots$ form an ascending sequence. Taking into account Lemma 11.10 we get

$$g_\lambda(\lim_{n\to\infty} A_n) = g_\lambda\left(\bigcup_{n=1}^\infty \bar{A}_n\right)$$

$$= \frac{1 - g_\lambda\left(\bigcup_{n=1}^\infty \bar{A}_n\right)}{1 + \lambda g_\lambda\left(\bigcup_{n=1}^\infty \bar{A}_n\right)}.$$

Taking into account the continuity by ascending sequences we get

$$g_\lambda(\lim_{n\to\infty} A_n) = \frac{1 - \lim_{n\to\infty} g_\lambda\left(\bar{A}_n\right)}{1 + \lambda \lim_{n\to\infty} g_\lambda\left(\bar{A}_n\right)}$$

$$= \lim_{n\to\infty} \frac{1 - g_\lambda\left(\bar{A}_n\right)}{1 + \lambda g_\lambda\left(\bar{A}_n\right)}.$$

Using Lemma 11.10, finally we obtain

$$g_\lambda(\lim_{n\to\infty} A_n) = \lim_{n\to\infty} \frac{1 - \frac{1 - g_\lambda(A_n)}{1 + \lambda g_\lambda(A_n)}}{1 + \lambda \frac{1 - g_\lambda(A_n)}{1 + \lambda g_\lambda(A_n)}}$$

$$= \lim_{n\to\infty} \frac{(1 + \lambda)g_\lambda(A_n)}{(1 + \lambda)} = \lim_{n\to\infty} g_\lambda(A_n).$$

■

**Proposition 11.12.** *Let $A, B \in \mathcal{B}$ not necessarily disjoint. Then we have*

$$g_\lambda(A \cup B) = \frac{g_\lambda(A) + g_\lambda(B) - g_\lambda(A \cap B) + \lambda \cdot g_\lambda(A) \cdot g_\lambda(B)}{1 + g_\lambda(A \cap B)}.$$

**Proof.** The proof is left as an exercise to the reader.

■

## 11.3   Possibility Measures

**Definition 11.13.** *(Dubois-Prade [51]) A possibility measure is a set function* $\Pi : \mathcal{B} \to [0,1]$ *such that*
  (i) $\Pi(\emptyset) = 0$, $\Pi(X) = 1$
  (ii) *If* $\{A_i | i \in I\} \subseteq \mathcal{B}$ *then*

$$\Pi \left( \bigcup_{i \in I} A_i \right) = \sup_{i \in I} \Pi(A_i).$$

**Definition 11.14.** *A possibility distribution is a fuzzy set* $\pi : X \to [0,1]$ *such that there exists* $x_0 \in X$ *with* $\pi(x_0) = 1$.

**Proposition 11.15.** *Given a possibility distribution* $\pi$, *for any* $A \in \mathcal{B}$ *we define* $\Pi(A) = \sup_{x \in A} \pi(x)$. *Then* $\Pi$ *is a possibility measure.*

**Proof.** (i) We adopt the convention $\Pi(\emptyset) = 0$. We have $\Pi(X) = \sup_{x \in X} \pi(x) = 1$.
  (ii) Let $\{A_i | i \in I\} \subseteq \mathcal{B}$. Then

$$\Pi \left( \bigcup_{i \in I} A_i \right) = \sup\{\pi(x) | x \in \bigcup_{i \in I} A_i\}$$

$$= \sup_{i \in I} \sup_{x \in A_i} \pi(x) = \sup_{i \in I} \Pi(A_i).$$

∎

**Proposition 11.16.** *If* $\Pi$ *is a possibility measure then* $\pi(x) = \Pi(\{x\})$ *is a possibility distribution.*

**Proof.** We have to prove that $\pi(x_0) = 1$ for some $x_0 \in X$. Since $\Pi(X) = 1$ and since we can write $X = \bigcup_{x \in X} \{x\}$ we obtain

$$\Pi(X) = \sup_{x \in X} \Pi(\{x\}) = 1$$

so there exists $x_0 \in X$ with $\Pi(\{x_0\}) = 1$, or equivalently $\pi(x_0) = 1$.     ∎

**Proposition 11.17.** *Let* $X$ *be a finite set and* $g : X \to [0,1]$ *be such that*
  (i) $g(\emptyset) = 0$, $g(X) = 1$
  (ii) *If* $A \cap B = \emptyset$ *then* $g(A \cup B) = \max\{g(A), g(B)\}$.
  *Then* $g$ *is a possibility measure.*

**Proof.** It is enough to prove that $g(A \cup B) = \max\{g(A), g(B)\}$ with $A, B$ not necessarily disjoint. For this aim we observe that

$$A \cup B = (A \setminus B) \cup (A \cap B) \cup (B \setminus A),$$

which is a union with disjoint components.

Condition (ii) implies

$$g(A \cup B) = \max\{g(A \setminus B), g(A \cap B), g(B \setminus A)\}$$

$$= \max\{\max\{g(A \setminus B), g(A \cap B)\}, \max\{g(A \cap B), g(B \setminus A)\}\}$$

$$= \max\{g(A), g(B)\}.$$

∎

The next theorem makes the connection between possibility measures and fuzzy measures.

**Theorem 11.18.** *A possibility measure that is continuous by descending sequences is a fuzzy measure.*

**Proof.** We can see that the condition $\Pi\left(\bigcup_{i \in I} A_i\right) = \sup_{i \in I} \Pi(A_i)$ implies monotonicity. Also, continuity by descending sequences is assumed. The only condition that needs to be proved is continuity by ascending sequences. Let $A_1 \subseteq A_2 \subseteq \ldots \subseteq A_n \subseteq \ldots$ be an ascending sequence. Then, since the limit of an increasing sequence coincides with its supremum we get

$$\Pi\left(\bigcup_{i=1}^{\infty} A_i\right) = \sup_{i=1,2,\ldots} \Pi(A_i) = \lim_{n \to \infty} \Pi(A_n).$$

∎

**Remark 11.19.** *Continuity by descending sequences is assumed, it is not a consequence of the properties of a possibility distribution in general. If $X$ is a finite set continuity by descending sequences is obtained, so the following corollary holds true.*

**Corollary 11.20.** *A possibility measure on a finite set is a fuzzy measure.*

**Proof.** Let $A_1 \supseteq A_2 \supseteq \ldots \supseteq A_n \supseteq \ldots$ be a descending sequence and $A = \bigcap_{i=1}^{\infty} A_i$. On a finite set we have $A = \bigcap_{i=1}^{\infty} A_i = A_k$ for some $k \in \{1, 2, \ldots\}$. Similar conclusion holds for ascending sequences. The rest of the proof is left to the reader as an exercise. ∎

## 11.4   Fuzzy Integrals

We will study in the present section fuzzy integrals, i.e., integrals obtained starting from the concepts of fuzzy measure. We will dedicate most of the section to Sugeno integrals and their equivalent definition due to Ralescu -Adams and we discuss the Choquet integral too.

**Definition 11.21.** *(Murofushi-Sugeno [113]) Let $\mu$ be a fuzzy measure. The Sugeno fuzzy integral of a positive measurable function $f : X \to [0, \infty)$ on a measurable subset $A \subseteq X$ is defined by*

$$(S) \int_A f \, d\mu = \bigvee_{\alpha \in [0, \infty)} \alpha \wedge \mu(F_\alpha \cap A),$$

*where*

$$F_\alpha = \{x \in X | f(x) \geq \alpha\}.$$

**Proposition 11.22.** *The Sugeno fuzzy integral has the following properties.*
  *a) If $A \subseteq B$ then*

$$(S) \int_A f \, d\mu \leq (S) \int_B f \, d\mu.$$

*b) If $\mu(A) = 0$ then*

$$(S) \int_A f \, d\mu = 0.$$

*c) If $c \geq 0$ is a constant then*

$$(S) \int_A c \, d\mu = c \wedge \mu(A).$$

*d) For $f, g : X \to [0, \infty)$, $f \leq g$ we have*

$$(S) \int_A f d\mu \leq (S) \int_B g \, d\mu.$$

**Proof.** The proof of a), b) and d) of the proposition are left to the reader as an exercise. Let us prove c) here. We observe that

$$F_\alpha = \{x \in X | c \geq \alpha\} = \left\{ \begin{array}{ll} \emptyset & \text{if} \quad c < \alpha \\ X & \text{if} \quad c \geq \alpha \end{array} \right. .$$

$$(S) \int_A c \, d\mu = \bigvee_{\alpha \in [0, \infty)} \alpha \wedge \mu(F_\alpha \cap A)$$

$$= \left( \bigvee_{\alpha \in [0, c)} \alpha \wedge 0 \right) \vee \bigvee_{\alpha \in [c, \infty)} \alpha \wedge \mu(A) = c \wedge \mu(A)$$

■

**Definition 11.23.** *(Ralescu-Adams, [124]) Let $\mu$ be a fuzzy measure and let*

$$s = \sum_{i=1}^{n} \alpha_i \chi_{A_i}$$

*be a simple function (i.e., piecewise constant with $A_i$ measurable). Further we denote by $S$ the set of simple functions on $X$. Let*

$$Q_A(s) = \bigvee_{i=1}^{n} \alpha_i \wedge \mu(A \cap A_i).$$

*The Ralescu-Adams fuzzy integral of a positive measurable function $f : X \to [0, \infty)$ on a subset $A \subseteq X$ is*

$$(RA) \int_A f \, d\mu = \sup_{s \in S} Q_A(s),$$

**Proposition 11.24.** *The Ralescu-Adams fuzzy integral given in this definition has the following properties.*
   *a) If $A \subseteq B$ then*

$$(RA) \int_A f \, d\mu \leq (RA) \int_B f \, d\mu.$$

*b) If $\mu(A) = 0$ then*

$$(RA) \int_A f \, d\mu = 0.$$

*c) If $c \geq 0$ is a constant then*

$$(RA) \int_A c \, d\mu = c \wedge \mu(A).$$

*d) For $f, g : X \to [0, \infty)$, $f \leq g$ we have*

$$(RA) \int_A f d\mu \leq (RA) \int_B g \, d\mu.$$

**Proof.** The proof of a), b) and d) of the proposition are left to the reader as an exercise. Let us prove c) here. We observe that we can write $s = c\chi_A$ which makes the integrand a simple function and then

$$Q_A(x) = c \wedge \mu(A)$$

and we obtain

$$(RA) \int_A c \, d\mu = c \wedge \mu(A).$$

■

Let us prove in the followings that the two integral concepts coincide.

**Theorem 11.25.** *(see e.g. Gal-Ban [70]) The Takagi-Sugeno and Ralescu-Adams integrals coincide and moreover*

$$(S)\int_X f \, d\mu = (RA)\int_X f \, d\mu = \sup_{A\in\mathcal{B}}[\mu(A) \wedge \inf_{x\in A} f(x)]$$

**Proof.** First we prove that

$$(RA)\int_X f \, d\mu = \sup_{A\in\mathcal{B}}[\mu(A) \wedge \inf_{x\in A} f(x)]$$

Let $A \in \mathcal{B}$ be measurable with $\inf_{x\in A} f(x) = \alpha_0$. Let us consider the simple function

$$s_0 = \alpha_0 \chi_A.$$

By the definition of $s_0$ we have $s_0 \leq f$ which allows us to write

$$(RA)\int_X f \, d\mu \geq (RA)\int_X s_0 \, d\mu \geq Q(s_0)$$

$$= \alpha_0 \wedge \mu(A) = \mu(A) \wedge \inf_{x\in A} f(x), \forall A \in \mathcal{B}.$$

We obtain

$$(RA)\int_X f \, d\mu \geq \sup_{A\in\mathcal{B}}[\mu(A) \wedge \inf_{x\in A} f(x)].$$

For the converse inequality we take

$$s = \sum_{i=1}^n \alpha_i \chi_{A_i} \leq f.$$

Then

$$Q(s) = \bigvee_{i=1}^n \alpha_i \wedge \mu(A_i).$$

Let us assume that the maximal element is attained at $i_0$ i.e.,

$$Q(s) = \alpha_{i_0} \wedge \mu(A_{i_0}).$$

Then we have

$$\alpha_{i_0} \leq \inf_{x\in A_{i_0}} f(x)$$

(otherwise we would not have $Q(s) \leq f(x)$).
It follows that

$$Q(s) \leq \mu(A_{i_0}) \wedge \inf_{x\in A_{i_0}} f(x) \leq \sup_{A\in\mathcal{B}}[\mu(A) \wedge \inf_{x\in A} f(x)]$$

and finally

$$(RA)\int_X f \, d\mu \leq \sup_{A\in\mathcal{B}}[\mu(A) \wedge \inf_{x\in A} f(x)].$$

Together with the symmetric case we obtain

$$(RA) \int_X f \, d\mu = \sup_{A \in \mathcal{B}} [\mu(A) \wedge \inf_{x \in A} f(x)].$$

For the Sugeno integral we take $\alpha_0 = \inf_{x \in A} f(x)$ then it is immediate that

$$A \subseteq F_{\alpha_0} = \{x \in X | f(x) \geq \alpha_0\}.$$

Then we get

$$(S) \int_X f \, d\mu = \bigvee_{\alpha \in [0,\infty)} \alpha \wedge \mu(F_\alpha \cap A)$$

$$\geq \alpha_0 \wedge \mu(A) = \mu(A) \wedge \inf_{x \in A} f(x)$$

and finally

$$(S) \int_X f \, d\mu \geq \sup_{A \in \mathcal{B}} [\mu(A) \wedge \inf_{x \in A} f(x)].$$

Since $F_\alpha$ is measurable we have $F_\alpha \in \mathcal{B}$. Then

$$(S) \int_A f \, d\mu = \bigvee_{\alpha \in [0,\infty)} \alpha \wedge \mu(F_\alpha) \leq \bigvee_{A \in \mathcal{B}} \alpha \wedge \mu(A).$$

$\blacksquare$

**Example 11.26.** *(see e.g. Gal-Ban [70]) Let $f : [0,2] \to [0,\infty)$, $f(x) = x$. Let $\mu$ be the Lebesgue measure. The Lebesgue measure is in particular a fuzzy measure too. Then*

$$\mu\{x | f(x) \geq \alpha\} = \mu\{x \in [0,2] | x \geq \alpha\} = 2 - \alpha$$

*and for the Sugeno integral we have*

$$(S) \int_A f \, d\mu = \bigvee_{\alpha \in [0,\infty)} \alpha \wedge \mu(F_\alpha)$$

$$= \bigvee_{\alpha \in [0,\infty)} \alpha \wedge (2 - \alpha) = 1.$$

**Definition 11.27.** *(see e.g., Sugeno-Murofushi [139]) The Choquet integral of a measurable function $f$ with respect to the fuzzy measure $\mu$ is defined as*

$$(C) \int f \, d\mu = \int_0^\infty \mu\{x | f(x) \geq \alpha\} d\alpha.$$

In what follows we give an equivalent definition.

**Definition 11.28.** *(see e.g., Sugeno-Murofushi [139]) Let* $0 = a_0 \leq a_1 \leq a_2 \leq ... \leq a_n$ *and* $A_i = \{x|f(x) \geq a_i\}$. *We have* $A_n \subseteq ... \subseteq A_1$. *We define the Choquet integral of a function of the form*

$$s = \sum_{i=1}^{n}(a_i - a_{i-1})\chi_{A_i} \in \mathcal{S}$$

*as*

$$(C)\int s \; d\mu = \sum_{i=1}^{n}(a_i - a_{i-1})\mu(A_i).$$

*For an arbitrary measurable function f we define*

$$(C)\int f \; d\mu = \sup\{(C)\int s \; d\mu | s \in \mathcal{S}, s \leq f\}.$$

**Proposition 11.29.** *(see e.g., Sugeno-Murofushi [139]) The concepts defined in the previous two definitions are equivalent i.e.,*

$$\int_0^{\infty} \mu\{x|f(x) \geq \alpha\}d\alpha = \sup\{(C)\int s \; d\mu | s \in \mathcal{S}, s \leq f\}.$$

**Proof.** We observe that

$$\sup\{(C)\int s \; d\mu | s \in \mathcal{S}, s \leq f\}$$

$$= \sup\left\{\sum_{i=1}^{n}\mu(A_i)(a_i - a_{i-1})|s \leq f\right\}$$

$$= \int_0^{\infty} \mu\{x|f(x) \geq \alpha\}d\alpha,$$

the later one being a Riemann integral. ∎

We list below some properties of the Choquet integral.

**Proposition 11.30.** *(see e.g., Sugeno-Murofushi [139])The Choquet integral has the following properties.*
   *1)* $\int 1_A \; d\mu = \mu(A)$.
   *2)* $\int af \; d\mu = a \int f \; d\mu$.
   *3) If* $f \leq g$ *then* $\int f \; d\mu \leq \int g \; d\mu$
   *4) If* $\mu \leq \nu$ *then* $\int f \; d\mu \leq \int f \; d\nu$
   *5) If* $\mu$ *is the Lebesgue measure then the Choquet integral coincides with a Lebesgue integral.*

**Proof.** The proofs of 1)-4) are left to the reader as an exercise. To prove 5) we observe that if $\mu$ is Lebesgue measure then as in Definition 11.28 we have

$$\mu(A_i) = \mu(A_{i+1}) + \mu(A_i - A_{i+1}).$$

Then

$$(C) \int s \, d\mu = \sum_{i=1}^{n} (a_i - a_{i-1}) \mu(A_i)$$

$$= \sum_{i=1}^{n} a_i (\mu(A_i) - \mu(A_{i+1}))$$

$$= \sum_{i=1}^{n} a_i \mu(A_i - A_{i+1}),$$

which converges to the Lebesgue integral and we get

$$(C) \int s \, d\mu = \int f \, d\mu.$$

∎

## 11.5   Problems

1. Prove Lemma 11.10.

2. Prove Proposition 11.12.

3. Prove Properties a), b), d) of Proposition 11.22.

4. Prove Properties a), b), d) of Proposition 11.24.

5. Prove Properties 1)-4) in Proposition 11.30.

6. Let $f : [0, 2] \to \mathbb{R}$, $f(x) = x^2$. Calculate the Sugeno integral

$$(S) \int_{[0,2]} f \, d\mu,$$

with $\mu$ being the Lebesgue measure.

# 12

# Fuzzy Clustering

## 12.1 Classical k-Means Clustering

Intuitively, clustering means partitioning a data set into clusters (subsets) whose objects share similar properties, i.e., they are near to each other in some well defined sense for "near" (see e.g., Jain-Dubes [82]). In this section we discuss the classical k-means clustering algorithm which is one of the basic classical clustering methods. It also stands at the basis of the subsequently described corresponding fuzzy techniques. Also, Faber [57] has proposed a continuous k-means clustering method. We discuss here a fuzzy version of Faber's algorithm the continuous fuzzy c-means method in Section 12.3. Those ideas are published here for the first time up to the author's best knowledge.

The classical k-means clustering (MacQueen [105]) is aimed to partition a data set into $k$ clusters (classical subsets), $k \in \mathbb{N}$, such that the data contained within each cluster is near to an average value.

More precisely the clustering problem can be formulated as an optimization question as follows.

**Clustering Problem.** Given $X = \{x_1, x_2, ..., x_n\} \subseteq \mathbb{R}^n$, a data set, and $k \leq n$ find $S_1, ..., S_k$ a partition of $X$ such that it minimizes

$$J((S_i)_{i=1,...,k}, (m_i)_{i=1,...,k}) = \sum_{i=1}^{k} \sum_{x_j \in S_i} \|x_j - m_i\|^2,$$

where $m_i$ is the mean of the points in $S_i$, $i = 1, ..., k$. We can simplify the notation above to $J(S_i, m_i)$, for convenience.

B. Bede: *Mathematics of Fuzzy Sets and Fuzzy Logic*, STUDFUZZ 295, pp. 213–219.
DOI: 10.1007/978-3-642-35221-8_12    © Springer-Verlag Berlin Heidelberg 2013

The following algorithm is called k-means algorithm and it is aimed to solve the above clustering problem. The algorithm consists of two steps.

**k-means algorithm.** (MacQueen [105])

Assign:

$$S_i = \{x_p | \, \|x_p - m_i\| \leq \|x_p - m_j\| \, , j = 1, ..., k\}$$

Update:

$$m_i = \frac{\sum_{x_j \in S_i} x_j}{|S_i|},$$

where $|S_i| = \sum_{x_j \in S_i} 1$ denotes the cardinality (number of elements) of the finite set.

**Proposition 12.1.** *The k-means algorithm converges in the sense that after a finite number of steps neither the Assignment nor the Update steps modify the output of the algorithm.*

**Proof.** It is easy to check that both the assignment and the update steps decrease the objective function. Indeed, if e.g., $S_i$ denotes the partition before update then we can denote

$$S_i' = \{x_p | \, \|x_p - m_i'\| \leq \|x_p - m_j'\| \, , j = 1, ..., k\}.$$

If a given value $x_q$ was previously "misplaced" in $S_k'$, i.e.,

$$\|x_q - m_k'\| \geq \|x_q - m_j'\| \, ,$$

for some $j \in \{1, ..., k\}$ then it will be placed in $S_j'$ and in this way $J(S_i, m_i)$ will decrease.

Let us analyze now the update step. We observe that if the sets $S_i$, $i = 1, ..., k$ are fixed we have

$$\frac{\partial J}{\partial m_i} = 2 \sum_{x_j \in S_i} (m_i - x_j), i = 1, ..., k$$

and also, $J$ is convex. Then a local minimum of $J$ is a global minimum and it is attained at its critical point i.e.,

$$m_i \sum_{x_j \in S_i} 1 = \sum_{x_j \in S_i} x_j$$

and finally

$$m_i = \frac{\sum_{x_j \in S_i} x_j}{\sum_{x_j \in S_i} 1}.$$

Since the point is a global minimum, the update step will decrease the value of $J(S_i, m_i)$. Finally since the search space for our problem is a finite set and since every step decreases the value of the functional $J(S_i, m_i)$, the algorithm will be convergent. ∎

## 12.2   Fuzzy c-Means

A crisp partition of a set does not allow partial membership degrees of a point in a cluster. Often it is convenient to have soft boundaries for clusters because e.g., a given point cannot be harshly categorized as belonging to a cluster or another. To allow partial membership to the clusters, these will need to be fuzzy sets. Based on this idea and based on the classical k-means algorithm, Bezdek proposed the fuzzy c-means algorithm described in the followings.

**Fuzzy Clustering Problem.** (Bezdek [28]) Given

$$X = \{x_1, x_2, ..., x_n\} \subseteq \mathbb{R}^n,$$

a data set, and $c \leq n$ and $m > 1$, find a fuzzy partition $u_1, ..., u_c$ of $X$, i.e., fuzzy sets on $X$ that fulfill the property

$$\sum_{i=1}^{c} u_{ik} = 1, \ u_{ik} = u_i(x_k)$$

such that the functional

$$J(u, v) = \sum_{i=1}^{c} \sum_{k=1}^{n} (u_{ik})^m \|x_k - v_i\|^2 ,$$

is minimized, where

$$v_i = \frac{\sum_{k=1}^{n} (u_{ik})^m x_k}{\sum_{k=1}^{n} (u_{ik})^m}.$$

is the center of the $i - th$ cluster, $i = 1, ..., c$.

The fuzzy c-means algorithm has two steps.

**Fuzzy c-means algorithm**
Assignment:

$$u_{ik} = \frac{1}{\sum_{j=1}^{c} \left( \frac{d_{ik}}{d_{jk}} \right)^{\frac{2}{m-1}}}, i = 1, ..., c$$

where

$$d_{ik} = \|x_k - v_i\|, i = 1, ..., c, k = 1, ..., n$$

and the norm is the Euclidean norm in $\mathbb{R}^n$ (however other norms could be also considered).

Update:

$$v_i = \frac{\sum_{k=1}^{n} (u_{ik})^m x_k}{\sum_{k=1}^{n} (u_{ik})^m}, i = 1, ..., c.$$

Considering the fuzzy clustering algorithm we can prove the following result.

**Theorem 12.2.** *(Bezdek [28]) The fuzzy c-means algorithm converges in the sense that $J^p(u,v)$ converges ($J^p$ denotes the value of the objective function in the p-th iteration).*

**Proof.** First we will consider the functional

$$J(u,v) = \sum_{i=1}^{c} \sum_{k=1}^{n} (u_{ik})^m \|x_k - v_i\|^2,$$

as a function of $u_{ik}$ with fixed $v_i$, and we prove that it has a minimum at

$$u_{ik} = \frac{1}{\sum_{j=1}^{c} \left(\frac{d_{ik}}{d_{jk}}\right)^{\frac{1}{m-1}}}, i = 1, ..., c.$$

Indeed if we consider the Lagrange function

$$L(u,v) = \sum_{i=1}^{c} \sum_{k=1}^{n} (u_{ik})^m \|x_k - v_i\|^2 + \sum_{k=1}^{n} \lambda_k \left(1 - \sum_{i=1}^{c} u_{ik}\right)$$

we obtain

$$\frac{\partial L}{\partial u_{ik}} = m(u_{ik})^{m-1} d_{ik}^2 - \lambda_k$$

with $d_{ik} = \|x_k - v_i\|$. Then

$$u_{ik} = \left(\frac{\lambda_k}{md_{ik}^2}\right)^{\frac{1}{m-1}}, i = 1, ..., c.$$

The fuzzy partition condition implies

$$\sum_{i=1}^{c} u_{ik} = \sum_{i=1}^{c} \left(\frac{\lambda_k}{md_{ik}^2}\right)^{\frac{1}{m-1}} = 1$$

and we obtain

$$\left(\frac{\lambda_k}{m}\right)^{\frac{1}{m-1}} = \frac{1}{\sum_{i=1}^{c} \left(\frac{1}{d_{ik}}\right)^{\frac{2}{m-1}}}$$

Then we obtain the critical point

$$u_{ik} = \frac{1}{\sum_{j=1}^{c} \left(\frac{d_{ik}}{d_{jk}}\right)^{\frac{2}{m-1}}}, i = 1, ..., c.$$

It is easy to check that the Hessian is positive semidefinite and so it is a convex function in its domain, and then $u_{ik}$ provided by the assignment step is a local minimum of $J(u,v)$, which is in particular a global minimum too, because of convexity. As a conclusion $J(u, v)$ is decreased in every assignment step.

Then, considering $J(u, v)$ as a function of $v_i$ with fixed $u_{ik}$ we can find a critical point of $J(u, v)$ from

$$\frac{\partial J}{\partial v_i} = -2 \sum_{k=1}^{n} (u_{ik})^m (x_k - v_i) = 0$$

as

$$v_i = \frac{\sum_{k=1}^{n} (u_{ik})^m x_k}{\sum_{k=1}^{n} (u_{ik})^m}, i = 1, ..., c.$$

Similar to the previous step the Hessian is positive semidefinite and so $J(u, v)$ with fixed $u$ has a local minimum at $v_i$, which is a global minimum too, and so the functional $J(u, v)$ is decreased in every update step.

Together the two steps of the algorithm will decrease the objective function, which ensures that the algorithm is convergent the objective function being bounded from below by 0.                                                    ∎

**Remark 12.3.** *Surely the above theorem ensures only the convergence of the algorithm without proving that it indeed converges to a local minimum of the functional $J(u, v)$. A deeper investigation of the fuzzy c-means algorithm is possible using Karush-Kuhn-Tucker theorems (see Jain-Dubes [82]).*

## 12.3   Continuous Fuzzy c-Means

Starting from Bezdek's fuzzy c-means [28] and Faber's continuous k-means [57] algorithms we can combine them into a continuous variant of the fuzzy c-means algorithm.

**Fuzzy Clustering Problem.** Given $\Omega \subseteq \mathbb{R}^n$, a region, and $c \in \mathbb{N}$, and $m > 1$, find a fuzzy partition $A_1, ..., A_c$ of $X$, i.e., fuzzy sets on $X$ that fulfill the property

$$\sum_{i=1}^{c} A_i(x) = 1,$$

such that the functional

$$J(u, v) = \int_{\Omega} \sum_{i=1}^{c} (A_i(x))^m \|x - v_i\|^2 \, dx,$$

is minimized, where

$$v_i = \frac{\int_{\Omega} (A_i(x))^m x \, dx}{\int_{\Omega} (A_i(x))^m \, dx}.$$

is the center of the $i - th$ cluster, $i = 1, ..., c.$

The continuous fuzzy c-means algorithm like its other variants will have two steps.

**Continuous fuzzy c-means algorithm**
Assignment:

$$A_i(x) = \frac{1}{\sum_{j=1}^{c} \left( \frac{\|x-v_i\|}{\|x-v_j\|} \right)^{\frac{2}{m-1}}}, i = 1, ..., c.$$

The norm can be any norm in $\mathbb{R}^n$ (however in the present discussion we use the Euclidean norm).

Update:

$$v_i = \frac{\int_{\Omega} (A_i(x))^m \, x dx}{\int_{\Omega} (A_i(x))^m}, i = 1, ..., c.$$

Considering the continuous fuzzy clustering algorithm we can prove the following result.

**Theorem 12.4.** *The continuous fuzzy c-means algorithm converges in the sense that $J^p(u,v)$ converges ($J^p$ denotes the value of the objective function in the p-th iteration).*

**Proof.** The approach presented here is not the most elegant, because it reduces the problem described here to the previous situation using a Riemann sum for approximating the integrals. First we will consider the functional

$$J(u,v) = \sum_{i=1}^{c} \int_{\Omega} (A_i(x))^m \|x - v_i\|^2 \, dx,$$

as a function of $A_i(x)$ with fixed $v_i$. We can approximate the functional $J(u,v)$ using a Riemann sum

$$J_n(u,v) = \sum_{i=1}^{c} \sum_{k=1}^{n} (A_i(x_k))^m \|x_k - v_i\|^2 (x_{k+1} - x_k).$$

and by the proof of Theorem 12.2 it has a minimum at

$$A_i(x_k) = \frac{1}{\sum_{j=1}^{c} \left( \frac{\|x_k-v_i\|}{\|x_k-v_j\|} \right)^{\frac{2}{m-1}}}, i = 1, ..., c.$$

If the assignment step calculates

$$A_i(x) = \frac{1}{\sum_{j=1}^{c} \left( \frac{\|x-v_i\|}{\|x-v_j\|} \right)^{\frac{2}{m-1}}}, i = 1, ..., c.$$

then $A_i(x_k)$ is automatically getting the value as described above, and so at the Assignment step the value of $J_n(u,v)$ is reduced, for any value of $n$.

$J(u, v)$ is within the approximation error of the Riemann sum, but it is still possible that it is increased. Let us assume that

$$|J(u, v) - J_n(u, v)| \leq \varepsilon.$$

Then

$$J_n(u, v) - \varepsilon \leq J(u, v) \leq J_n(u, v) + \varepsilon,$$

i.e., however $J(u, v)$ may increase, at an individual step, its approximation $J_n(u, v)$ is decreased at the assignment step.

Then, considering $J(u, v)$ as a function of $v_i$ with fixed fuzzy partitions we can find a critical point of $J(u, v)$ from

$$\frac{\partial J}{\partial v_i} = -2 \int_\Omega (A_i(x))^m (x - v_i) \, dx = 0$$

and we obtain

$$v_i = \frac{\int_\Omega (A_i(x))^m x dx}{\int_\Omega (A_i(x))^m dx}, i = 1, ..., c.$$

Similar to the previous step the Hessian is positive semidefinite and so $J(u, v)$ has a local minimum at $v_i$, which is a global minimum too, and so the functional $J(u, v)$ is decreased in every update step. Same can be said in particular about $J_n$, which also decreases at every update step.

Taking $\varepsilon = \frac{1}{n}$ we obtain that in the assignment step we have

$$J_n(u, v) - \frac{1}{n} \leq J(u, v) \leq J_n(u, v) + \frac{1}{n},$$

so $J(u, v)$ is bounded by the sequence $J_n(u, v) - \frac{1}{n}$ and $J_n(u, v) + \frac{1}{n}$ and so the algorithm converges.    ∎

## 12.4 Problems

1. Let us consider the functional

$$J(u, v) = \sum_{i=1}^{c} \sum_{k=1}^{n} e^{u_{ik}} \|x_k - v_i\|^2.$$

Describe a fuzzy c-means algorithm that should minimize the given functional.

2. Prove that the algorithm described in Problem 1 is convergent.

3. Investigate generalizations of the fuzzy c-means algorithms similar to Problem 1.

# 13

# Fuzzy Transform

## 13.1 Definition of Fuzzy Transforms

Fuzzy Transform was proposed in Perfilieva in [121] and Perfilieva [120] and it is an approximation method based on fuzzy sets. Also, it can be seen as a fuzzy set-based analogue of the Fourier Transform. In the present chapter we follow in great lines the presentation and discussion in Bede-Rudas [23].

**Definition 13.1.** *A fuzzy partition of an interval $[a, b]$ is defined as a family $\mathbb{A} = \{A_1, A_2, ..., A_k\}$ of fuzzy sets on $[a, b]$, $A_i : [a, b] \to [0, 1], i = 1, ..., k$ such that*

$$\sum_{i=1}^{k} A_i(x) = 1, \forall x \in [a, b].$$

*The fuzzy sets $A_i$ are called the atoms of the partition.*

**Example 13.2.** *We consider a classical partition $a = y_0 = y_1 \leq ... \leq y_k = y_{k+1} = b$ of the interval $[a, b]$, with triangular fuzzy numbers*

$$A_i = (x_{i-1}, x_i, x_{i+1}), i = 1, ..., k.$$

*These fuzzy sets form a fuzzy partition (see Fig. 13.1).*

If $a = y_0 \leq y_1 < ... < y_k \leq y_{k+1} = b$ is a classical partition of the interval $[a, b]$ and let $\delta = \max_{i=0,...,k} |y_{i+1} - y_i|$ be the norm of the partition.

In what follows we present the Fuzzy Transform (F-transform) proposed in Perfilieva [121].

B. Bede: *Mathematics of Fuzzy Sets and Fuzzy Logic*, STUDFUZZ 295, pp. 221–246.
DOI: 10.1007/978-3-642-35221-8_13      © Springer-Verlag Berlin Heidelberg 2013

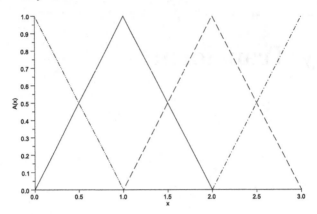

**Fig. 13.1** A fuzzy partition with triangular atoms

**Definition 13.3.** *(Perfilieva [121]) We consider a fuzzy partition* $\mathbb{A}$. *The* **continuous F-transform** *is given by*

$$f_i = \frac{\int_a^b A_i(x) f(x) dx}{\int_a^b A_i(x) dx}, \qquad (13.1)$$

$i = 1, ..., k.$

*The* **discrete F-transform** *is given by*

$$f_i = \frac{\sum_{j=1}^n A_i(x_j) f(x_j)}{\sum_{j=1}^n A_i(x_j)}, \qquad (13.2)$$

*where* $x_j \in I$, $j = 1, ..., n$ *are given data* $(n \geq 1)$ *such that for each* $i \in \{1, ..., k\}$ *there exists* $p \in \{1, ..., n\}$ *with* $x_p \in supp A_i$, *i.e. in the support of each atom of the partition we find at least one data point.*

where $f_i$ are the components of the continuous F-transform.

**Definition 13.4.** *(Perfilieva [121]) The* **continuous inverse F-transform** *is*

$$F_k(x) = \sum_{i=1}^k A_i(x) f_i. \qquad (13.3)$$

*where* $f_i$ *are the components of the continuous F-transform, and the* **discrete inverse F-transform** *is*

$$F_{n,k}(x) = \sum_{i=1}^k A_i(x) f_i. \qquad (13.4)$$

where $f_i$ are the components of the discrete F-transform.

Positive linear operators are playing a special role in our investigation.

**Definition 13.5.** *For the discrete case let $\mathcal{F}_{n,k} : C([a,b]) \to C([a,b])$ be given by*

$$\mathcal{F}_{n,k}(f)(x) = F_{n,k}(x),$$

*where $F_{n,k}(x)$ is given by (13.4) and $f_i$ given by (13.2). The operator $\mathcal{F}_{n,k}$ is called the discrete inverse F-transform approximation operator.*

*If we consider the composition between the inverse F-transform and the continuous F-transform we take $\mathcal{F}_k : C([a,b]) \to C([a,b])$, given by*

$$\mathcal{F}_k(f)(x) = F_k(x),$$

*where $F_k(x)$ is given by (13.3) and $f_i$ given by (13.1). The operator $\mathcal{F}_k$ is called the continuous inverse F-transform approximation operator.*

**Example 13.6.** *Let us consider the fuzzy partition with four triangles in Fig. 13.1*

$$A_1 = (0,0,1), \ A_2 = (0,1,2), \ A_3 = (1,2,3), \ A_4 = (2,3,3)$$

*Let us calculate the components of the continuous F-transform associated to the function $f(x) = x^2$.*

$$f_1 = \frac{\int_0^1 (1-x)x^2 dx}{\int_0^1 (1-x)dx} = \frac{1}{6},$$

$$f_2 = \frac{\int_0^1 xx^2 dx + \int_1^2 (2-x)x^2 dx}{\int_0^1 xdx + \int_0^1 (2-x)dx} = \frac{7}{6},$$

$$f_3 = \frac{\int_1^2 (x-1)x^2 dx + \int_2^3 (3-x)x^2 dx}{\int_1^2 (x-1)dx + \int_2^3 (3-x)dx} = \frac{25}{6},$$

$$f_4 = \frac{\int_2^3 (x-2)x^2 dx}{\int_2^3 (x-2)dx} = \frac{43}{6}.$$

*Then the inverse F-transform becomes*

$$\mathcal{F}_k(f)(x) = F_k(x) = A_1(x) \cdot \frac{1}{6} + A_2(x) \cdot \frac{7}{6} + A_3(x) \cdot \frac{25}{6} + A_4(x) \cdot \frac{43}{6}.$$

*The direct and inverse continuous F-transform are illustrated in Fig. 13.2. The more practical discrete F-transform can be numerically calculated.*

**Example 13.7.** *Let us consider a fuzzy partition based on trigonometric functions. Let $a = y_0 \leq \ldots \leq y_n = b$ be a partition with equally spaces knots $y_i = a + h \cdot i$, with $h = \frac{b-a}{n}$, $i = 0, \ldots, n$. Let us consider $y_{-1} \leq a$, $b \leq y_{n+1}$ two auxiliary knots. Let us take (see Fig. 13.3)*

$$A_i(x) = \begin{cases} \frac{1}{2}\left(\cos\frac{\pi}{h}(x - y_i) + 1\right) & \text{if} \quad x \in [y_{i-1}, y_{i+1}] \\ 0 & \text{otherwise} \end{cases} \quad i = 0, \ldots, n.$$

*Let us calculate the discrete F-transform associated to the function $f(x) = x^2$. The direct and inverse discrete F-transform are illustrated in Fig. 13.4.*

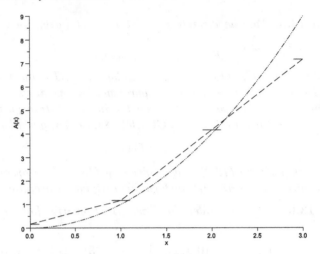

**Fig. 13.2** Direct and inverse continuous fuzzy transform

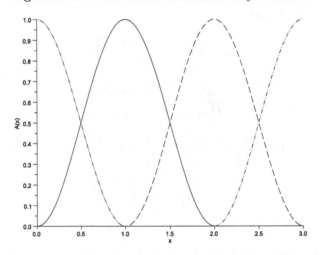

**Fig. 13.3** Fuzzy Partition with trigonometric membership functions

**Proposition 13.8.** $\mathcal{F}_{n,k} \in \mathcal{L}(C[a,b])$ and $\mathcal{F}_k \in \mathcal{L}(C[a,b])$ are positive linear operators.

**Proof.** The proof is left to the reader as an exercise. ■

Also, we can regard each of the direct F-transforms $\mathbf{f}_i$ as a linear form $\mathbf{f}_i :$ $C([a,b]) \to \mathbb{R}$, $\mathbf{f}_i(f) = f_i$ as in (13.1) or (13.2).

An interesting result concerning F-transforms is their optimality property obtained in Perfilieva [121].

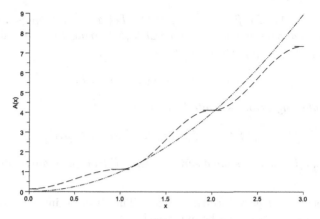

**Fig. 13.4** Direct and inverse discrete fuzzy transform

**Proposition 13.9.** *For each $i = 1, ..., k$, the point $y = f_i$, (with $f_i$ being a component of the continuous F-transform) minimizes in terms of $y$ the function*

$$\Phi(y) = \int_a^b (f(x) - y)^2 A_i(x) dx.$$

**Proof.** We observe that a minimizer has to be a critical point of the convex function $\Phi$, i.e.,

$$\Phi'(y) = -2 \int_a^b (f(x) - y) A_i(x) dx = 0,$$

and we obtain that the critical point is indeed the F-transform

$$y = \frac{\int_a^b A_i(x) f(x) dx}{\int_a^b A_i(x) dx} = f_i, i = 1, ..., k.$$

∎

**Proposition 13.10.** *The component $f_i$ of the discrete F-transform minimizes in terms of $y$ the function*

$$\Phi(y) = \sum_{j=1}^n (f(x_j) - y)^2 A_i(x_j).$$

**Proof.** The proof is left to the reader as an exercise.        ∎

## 13.2    Error Estimates for F-Transforms Based on Fuzzy Partitions with Small Support

The original definition of the Fuzzy transform assumes a fuzzy partition with small support. In this case the uniform convergence of the composition of the inverse and direct F-transform was shown in Perfilieva [121]. We consider a generalization proposed in Bede-Rudas [23].

**Theorem 13.11.** *(Perfilieva [121], [120]) Let* $a = y_0 \leq y_1 \leq ... \leq y_k \leq y_{k+1} = b$ *be a partition of the interval* $[a, b]$ *having the norm* $\delta$. *If* $\mathbb{A} = \{A_1, ..., A_k\}$ *is a fuzzy partition with*

$$(A_i)_0 \subseteq [y_{i-1}, y_{i+1}], i = 1, ..., k,$$

*then the following error estimate holds true*

$$|\mathcal{F}_{n,k}(f)(x) - f(x)| \leq 2\omega(f, \delta), \forall x \in [a, b],$$

*where* $\mathcal{F}_{n,k}(f)(x)$ *denotes the discrete inverse F-transform approximation operator.*

**Proof.** Let $x \in [a, b]$ be arbitrary fixed. The discrete inverse F-transform approximation operator can be expressed as

$$\mathcal{F}_{n,k}(f)(x) = \sum_{i=1}^{k} A_i(x) \frac{\sum_{j=1}^{n} A_i(x_j) f(x_j)}{\sum_{j=1}^{n} A_i(x_j)}$$

and $f(x)$ can be written as

$$f(x) = \sum_{i=1}^{k} A_i(x) \frac{\sum_{j=1}^{n} A_i(x_j) f(x)}{\sum_{j=1}^{n} A_i(x_j)}.$$

Then we get

$$|\mathcal{F}_{n,k}(f)(x) - f(x)|$$

$$\leq \sum_{i=1}^{k} A_i(x) \frac{\sum_{j=1}^{n} A_i(x_j) |f(x) - f(x_j)|}{\sum_{j=1}^{n} A_i(x_j)},$$

and from Theorem A.39 we have

$$|\mathcal{F}_{n,k}(f)(x) - f(x)| \leq \sum_{i=1}^{k} \frac{\sum_{j=1}^{n} A_i(x) A_i(x_j) \omega(f, |x - x_j|)}{\sum_{j=1}^{n} A_i(x_j)}.$$

We observe that the product $A_i(x) A_i(x_j) \neq 0$ only when $x, x_j$ are in the same interval $[y_{i-1}, y_{i+1}]$. In this case we have $|x - x_j| \leq 2\delta$ and we get

$$|\mathcal{F}_{n,k}(f)(x) - f(x)| \leq 2 \sum_{i=1}^{k} \frac{\sum_{j=1}^{n} A_i(x) A_i(x_j) \omega(f, \delta)}{\sum_{j=1}^{n} A_i(x_j)}$$

and finally

$$|\mathcal{F}_{n,k}(f)(x) - f(x)| \leq 2\omega(f, \delta).$$

∎

In the next theorem we obtain an error estimate in the context of a fuzzy partition with small support. This below result slightly extends those in Perfilieva [121].

**Theorem 13.12.** *Let* $a = y_{1-r} \leq y_1 \leq ... \leq y_k \leq y_{k+r} = b$ *be a partition of the interval* $[a, b]$ *having the norm* $\delta$. *If* $\mathbb{A} = \{A_1, ..., A_k\}$ *is a fuzzy partition with supports*

$$(A_i)_0 \subseteq [y_{i-r}, y_{i+r}], i = 1, ..., k$$

*then the following error estimate holds true*

$$|\mathcal{F}_{n,k}(f)(x) - f(x)| \leq 2r\omega(f, \delta), \forall x \in [a, b],$$

*where* $\mathcal{F}_{n,k}(f)(x)$ *denotes the discrete inverse F-transform approximation operator.*

**Proof.** The proof is left as an exercise.   ∎

The following Corollary gives the uniform convergence of the composition of the direct and inverse discrete F-transforms.

**Corollary 13.13.** *If* $f$ *is continuous, then there exists a sequence of fuzzy partitions* $\mathbb{A}_m$ *such that the composition of the inverse and direct discrete F-transform converges uniformly to* $f$.

Similar results hold for the case of continuous fuzzy transforms.

**Theorem 13.14.** *Let* $a = y_0 \leq y_1 \leq ... \leq y_k \leq y_{k+1} = b$ *be a partition of the interval* $[a, b]$ *having the norm* $\delta$. *If* $\mathbb{A} = \{A_1, ..., A_k\}$ *is a fuzzy partition with*

$$(A_i)_0 \subseteq [y_{i-1}, y_{i+1}], i = 1, ..., k$$

*then the following error estimate holds true*

$$|\mathcal{F}_k(f)(x) - f(x)| \leq 2\omega(f, \delta), \forall x \in [a, b],$$

*where* $\mathcal{F}_k(f)(x)$ *denotes the continuous inverse F-transform approximation operator.*

**Proof.** Similar to the proof of Theorem 13.11 we have

$$|\mathcal{F}_k(f)(x) - f(x)| \leq \sum_{i=1}^{k} \frac{\int_a^b A_i(x)A_i(t)\omega(f, |x - t|)dt}{\int_a^b A_i(t)dt}$$

Since the product $A_i(x)A_i(t) \neq 0$ if and only if $x, t \in [y_{i-1}, y_{i+1}]$, and in this interval we have $|x - t| \leq 2\delta$ we get

$$|\mathcal{F}_k(f)(x) - f(x)| \leq 2 \sum_{i=1}^{k} \frac{\int_{y_{i-1}}^{y_{i+1}} A_i(x)A_i(t)dt}{\int_{y_i}^{y_{i+r}} A_i(t)dt} \omega(f, \delta)$$

$$\leq 2\omega(f, \delta).$$

   ∎

**Theorem 13.15.** *Let* $a = y_0 \leq y_1 \leq ... \leq y_k \leq y_{k+1} = b$ *be a partition of the interval* $[a, b]$ *having the norm* $\delta$. *If* $\mathbb{A} = \{A_1, ..., A_k\}$ *is a fuzzy partition with*

$$(A_i)_0 \subseteq [y_{i-r}, y_{i+r}], i = 1, ..., k$$

*then the following error estimate holds true*

$$|\mathcal{F}_k(f)(x) - f(x)| \leq 2r\omega(f, \delta), \forall x \in [a, b],$$

*where* $\mathcal{F}_k(f)(x)$ *denotes the continuous inverse F-transform approximation operator.*

**Proof.** The proof is immediate, being similar to the proof of the previous theorem. ∎

**Corollary 13.16.** *If* $f$ *is continuous, then there exists a sequence of fuzzy partitions* $\mathbb{A}_m$ *such that the composition of the inverse and direct continuous F-transform converges uniformly to* $f$.

## 13.3  B-Splines Based F-Transform

Let us recall the definitions of the classical B-splines of order $r - 1$ with $r \geq 1$. Let $t_0 \leq t_1 \leq .... \leq t_r$ be points in $\mathbb{R}$, with $t_r \neq t_0$. The B-spline $M$ is given by

$$M(x) = M(x; t_0, ..., t_r) = r[t_0, ..., t_r](\cdot - x)_+^{r-1},$$

where for fixed $x$, the notation $(\cdot - x)_+$ denotes the function

$$(\cdot - x)_+(t) = (t - x)_+ = \max\{0, t - x\}$$

and $[t_0, ..., t_r]f$ denotes the divided difference of $f$ (see e.g. DeVore- Lorentz [43]) defined recursively as

$$[t_k]f = f(t_k), k = 0, ..., r$$

$$[t_i, ..., t_j]f = \frac{[t_{i+1}, ..., t_j]f - [t_i, ..., t_{j-1}]f}{t_j - t_i}, 0 \leq i < j \leq r.$$

The B-spline $N$ is defined by

$$N(x; t_0, ..., t_r) = \frac{1}{r}(t_r - t_0)M(x; t_0, ..., t_r). \tag{13.5}$$

Let $t_0 \leq t_1 \leq ... \leq t_n$ be a sequence of points in a given interval $[a, b]$, called basic knots. We need some auxiliary knots $t_{-r+1} \leq ... \leq t_0 = a$ and $b = t_{n+1} \leq ... \leq t_{n+r}$. To a given sequence of knots corresponds a sequence of crisp B-splines

$$N_i^r(x) = N(x; t_i, ..., t_{i+r}),$$

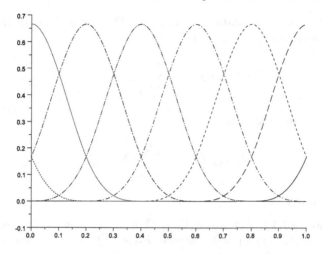

**Fig. 13.5** Cubic B-spline based fuzzy partition

for $i \in \{-r+1, ..., n\}$. The following identity holds

$$\sum_{i=-r+1}^{n} N_i^r(x) = 1, \tag{13.6}$$

i.e., the B-splines form a fuzzy partition. Cox-de Boor recursion formula (see DeVore- Lorentz [43]) is important both from practical and theoretical point of view

$$N_i^1(x) = \begin{cases} 1 \text{ if } t_i \le x < t_{i+1} \\ 0 \text{ otherwise} \end{cases}$$

$$N_i^r(x) = \frac{x - t_i}{t_{i+r-1} - t_i} N_i^{r-1}(x) + \frac{t_{i+r} - x}{t_{i+r} - t_{i+1}} N_{i+1}^{r-1}(x),$$

for any $r \ge 2, i = -r+1, ..., n$.

If we consider an equidistant knot sequence, B-splines form a fuzzy partition as shown in Fig. 13.5.

Based on the results of the previous section we consider the B-spline based fuzzy transform proposed in Bede-Rudas [23]. The B-spline-based continuous F-transform is given by

$$f_i = \frac{\int_a^b N_i^r(x) f(x) dx}{\int_a^b N_i^r(x) dx}, \quad i = -r+1, ..., n$$

and the discrete F-transform is

$$f_i = \frac{\sum_{j=1}^{m} N_i^r(x_j) f(x_j)}{\sum_{j=1}^{m} N_i^r(x_j)}, \quad i = -r+1, ..., n$$

where $x_j \in I$, are given data $(j = 1, ..., m)$. The composition of the inverse and direct F-transform for $x \in [a, b]$ is

$$\mathcal{F}_n(f)(x) = \sum_{i=-r+1}^{n} N_i^r(x) f_i,$$

for the continuous case, and

$$\mathcal{F}_{n,k}(f)(x) = \sum_{i=-r+1}^{n} N_i^r(x) f_i,$$

for the discrete case.

**Proposition 13.17.** *Let $f : [a, b] \to \mathbb{R}$ be a continuous function and let $\mathcal{F}_n(f)(x)$ and $\mathcal{F}_{n,k}(f)(x)$ be the inverse B-spline based F-transform approximation operators. Then we have*

$$|\mathcal{F}_n(f)(x) - f(x)| \leq r\omega(f, \delta) \text{ and} \qquad (13.7)$$
$$|\mathcal{F}_{n,k}(f)(x) - f(x)| \leq r\omega(f, \delta)$$

*where $r - 1$ is the order of the spline considered, and $\delta = \max_{i=0}^{n} |t_{i+1} - t_i|$.*

**Proof.** The proof is similar to that of Theorem 13.12 and it is left to the reader as exercise. ∎

The estimate in (13.7) can be generally improved, but if $r$ is small compared to $n$, we have a sufficiently good estimate. In applications, to keep the problem's simplicity one needs to choose a relatively small value for $r$. Indeed, the most popular applications of classical B-splines use cubic splines.

**Example 13.18.** *We will take F-transforms and inverse F-transforms of the functions $f_1, f_2 : [0, 1] \to \mathbb{R}_+$*

$$f_1(x) = 2 + \sin \frac{1}{x + 0.15} \qquad (13.8)$$

*and*

$$f_2(x) = \begin{cases} 1 & \text{if} \quad 0 \leq x \leq 0.4 \\ 10x - 3 & \text{if} \quad 0.4 < x \leq 0.5 \\ 2 & \text{if} \quad 0.5 < x \leq 1 \end{cases} . \qquad (13.9)$$

*B-splines of different orders can be used to approximate these functions on the $[0, 1]$ interval. We use cubic B-splines here. In Fig.13.6 we show the F-transform approximation for the function $f_1$.*

*In Fig. 13.7, cubic B-spline based F-transform is considered for the function $f_2$.*

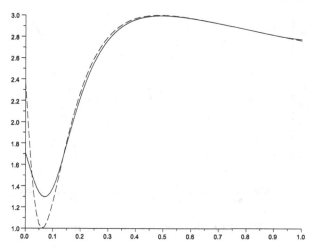

**Fig. 13.6** Approximation by cubic B-spline F-transform. Dashed line $= f_1$, solid line $=$ inverse F-transform

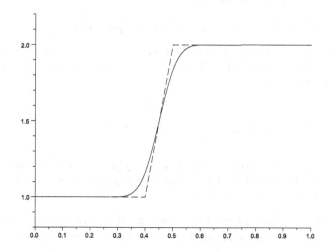

**Fig. 13.7** Approximation by cubic B-spline F-transform. Solid line $= f_2$ Dashed line $=$ inverse F-transform

## 13.4 Shepard Kernels Based F-Transform

We consider in what follows a Shepard-like fuzzy partition. This was proposed in Bede-Rudas [23].

**Definition 13.19.** *Consider $y_0 < \ldots < y_k \in [a, b]$. We say that a fuzzy partition $\mathbb{A} = \{A_i\}_{i \in \{0,\ldots,k\}}$ determines the Shepard-like fuzzy partition if it is of the following form*

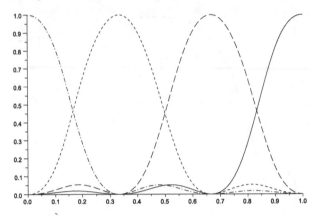

**Fig. 13.8** Shepard type fuzzy partition

$$A_i(x) = \begin{cases} \frac{|x-y_i|^{-\lambda}}{\sum_{j=0}^k |x-y_j|^{-\lambda}}, & \text{if } x \in [a,b] \setminus \{y_0, ..., y_k\}, \\ \delta_{ij}, & \text{otherwise,} \end{cases} \tag{13.10}$$

where $\delta_{ij}$ is Kronecker's delta, and $\lambda > 2$ is a parameter.

We consider the interval $[0,1]$ and the points $y_1, ..., y_k$ are considered equally spaced. Now, let us recall that if $z_i$, $i = 1, ..., n$ are equally-spaced points in the unit interval then

$$\left( \sum_{i=0}^n |x - z_i|^{-\gamma} \right)^{-1} = \mathcal{O}(n^{-\gamma}), \tag{13.11}$$

for any $\gamma > 1$ (see Szabados [142]), where $f(h) = \mathcal{O}(g(h))$ has the usual interpretation, i.e., there exist $C, x_0$ such that $f(x) \leq Cg(x)$ for any $x \geq x_0$. Let us recall the following lemma:

**Lemma 13.20.** *(Xiao-Zhou [152], Lemma 1)For any $x \in [0,1]$ and $\gamma > 1$, if $z_i$, $i = 1, ..., n$ are equally-spaced points in the unit interval then*

$$\frac{|x - z_k|^{-\gamma}}{\sum_{i=0}^n |x - z_i|^{-\gamma}} \leq \mathcal{O}\left( \left| \lfloor (n+1)x \rfloor - k \right| + 1 \right)^{-\gamma}, \tag{13.12}$$

where $\lfloor x \rfloor$ is the greatest integer not exceeding $x$.

The following theorem provides an error estimate for the inverse F-transform based on Shepard kernels.

**Theorem 13.21.** *(Bede-Rudas [23]) Let $y_0, ..., y_k$ and $x_0, ..., x_n$ be equally-spaced points in $[0,1]$. If $A_i(x)$ denote Shepard kernels (13.10) having the* *parameter $\lambda > 2$, then for the discrete inverse F-transforms approximation operator $\mathcal{F}_{n,k}(f)(x)$ given by (13.3) and (13.2) there exists an absolute constant $C$ such that the following error estimate*

$$|\mathcal{F}_{n,k}(f)(x) - f(x)| \leq C\omega\left(f, \frac{1}{m}\right),$$

*holds, with $m = \min\{n, k\}$.*

**Proof.** As in the proof of the previous theorem we have

$$|\mathcal{F}_{n,k}(f)(x) - f(x)| \leq \sum_{i=0}^{k} \frac{\sum_{j=0}^{n} A_i(x) A_i(x_j) \omega(f, |x - x_j|)}{\sum_{j=0}^{n} A_i(x_j)}.$$

The following inequality is obtained from properties of the modulus of continuity:

$$|\mathcal{F}_{n,k}(f)(x) - f(x)| \tag{13.13}$$

$$\leq \left(1 + m \sum_{i=0}^{k} A_i(x) \frac{\sum_{j=0}^{n} A_i(x_j) |x - x_j|}{\sum_{j=0}^{n} A_i(x_j)}\right) \omega\left(f, \frac{1}{m}\right).$$

We have to estimate the expression

$$E_{n,k}(x) = \sum_{i=0}^{k} A_i(x) \frac{\sum_{j=0}^{n} A_i(x_j) |x - x_j|}{\sum_{j=0}^{n} A_i(x_j)}$$

$$= \sum_{i=0}^{k} \frac{|x - y_i|^{-\lambda}}{\sum_{l=0}^{k} |x - y_l|^{-\lambda}} \frac{\sum_{j=0}^{n} \frac{|x_j - y_i|^{-\lambda}}{\sum_{l=0}^{k} |x_j - y_l|^{-\lambda}} |x - x_j|}{\sum_{j=0}^{n} \frac{|x_j - y_i|^{-\lambda}}{\sum_{l=0}^{k} |x_j - y_l|^{-\lambda}}}.$$

We observe that

$$E_{n,k}(x) = \sum_{i=0}^{k} \frac{|x - y_i|^{-\lambda}}{\sum_{l=0}^{k} |x - y_l|^{-\lambda}} \sum_{j=0}^{n} \frac{|x_j - y_i|^{-\lambda} |x - x_j|}{\sum_{j=0}^{n} |x_j - y_i|^{-\lambda}}$$

$$\leq \sum_{i=0}^{k} \frac{|x - y_i|^{-\lambda}}{\sum_{l=0}^{k} |x - y_l|^{-\lambda}} \sum_{j=0}^{n} \frac{|x_j - y_i|^{-\lambda} (|x - y_i| + |x_j - y_i|)}{\sum_{j=0}^{n} |x_j - y_i|^{-\lambda}}$$

$$= \sum_{i=0}^{k} \frac{|x - y_i|^{-\lambda}}{\sum_{l=0}^{k} |x - y_l|^{-\lambda}} |x - y_i| \sum_{j=0}^{n} \frac{|x_j - y_i|^{-\lambda}}{\sum_{j=0}^{n} |x_j - y_i|^{-\lambda}}$$

$$+ \sum_{i=0}^{k} \frac{|x - y_i|^{-\lambda}}{\sum_{l=0}^{k} |x - y_l|^{-\lambda}} \sum_{j=0}^{n} \frac{|x_j - y_i|^{-\lambda}}{\sum_{j=0}^{n} |x_j - y_i|^{-\lambda}} |x_j - y_i|.$$

From Lemma 13.20, for the equally spaced points $y_0, ..., y_k$, we have

$$\frac{|x - y_i|^{-\lambda}}{\sum_{l=0}^{k} |x - y_l|^{-\lambda}} = \mathcal{O}\left(|[(k+1)x] - i| + 1\right)^{-\lambda},$$

and for the equally spaced points $x_0, ..., x_n$ we obtain

$$\frac{|x_j - y_i|^{-\lambda}}{\sum_{j=0}^{n} |x_j - y_i|^{-\lambda}} = \mathcal{O}\left(|[(n+1)y_i] - j| + 1\right)^{-\lambda},$$

and it follows

$$E_{n,k}(x) = \mathcal{O}\left(\sum_{i=0}^{k} \left(|[(k+1)x] - i| + 1\right)^{-\lambda} |x - i/k|\right.$$

$$\left. \cdot \sum_{j=0}^{n} \left(|[(n+1)y_i] - j| + 1\right)^{-\lambda}\right)$$

$$+ \mathcal{O}\left(\sum_{i=0}^{k} \left(|[(k+1)x] - i| + 1\right)^{-\lambda} \cdot \sum_{j=0}^{n} \left(|[(n+1)y_i] - j| + 1\right)^{-\lambda} |j/n - y_i|\right)$$

$$= \mathcal{O}\left(\frac{1}{k} \sum_{i=0}^{k} \left(|[(k+1)x] - i| + 1\right)^{-\lambda} |kx - i| \cdot \sum_{j=0}^{n} \left(|[(n+1)y_i] - j| + 1\right)^{-\lambda}\right)$$

$$+ \mathcal{O}\left(\frac{1}{n} \sum_{i=0}^{k} \left(|[(k+1)x] - i| + 1\right)^{-\lambda} \cdot \sum_{j=0}^{n} \left(|[(n+1)y_i] - j| + 1\right)^{-\lambda} |ny_i - j|\right).$$

Now let us observe that $|kx - i| \le |[(k+1)x] - i| + 1$. Indeed, if $kx \ge i$ we have

$$|kx - i| = kx - i \le (k+1)x - i$$

$$\le |[(k+1)x] - i + 1| \le |[(k+1)x] - i| + 1.$$

Similarly, if $kx < i$ we obtain

$$|kx - i| = i - kx \le i - (k+1)x$$

$$\le |[(k+1)x] - i| \le |[(k+1)x] - i| + 1.$$

Also, we have

$$|ny_i - j| \le |[(n+1)y_i] - j| + 1$$

and we obtain

$$E_{n,k}(x) = \mathcal{O}\left(\frac{1}{k}\sum_{i=0}^{k}(|[(k+1)x]-i|+1)^{-\lambda+1}\cdot\sum_{j=0}^{n}(|[(n+1)y_i]-j|+1)^{-\lambda}\right)$$

$$+\mathcal{O}\left(\frac{1}{n}\sum_{i=0}^{k}(|[(k+1)x]-i|+1)^{-\lambda}\cdot\sum_{j=0}^{n}(|[(n+1)y_i]-j|+1)^{-\lambda+1}\right).$$

We observe that

$$E_{n,k}(x) = \mathcal{O}\left(\frac{1}{k}\sum_{\alpha=1}^{\infty}\alpha^{-\lambda+1}\sum_{\beta_\alpha=1}^{\infty}\beta_\alpha^{-\lambda}\right)$$

$$+\mathcal{O}\left(\frac{1}{n}\sum_{\alpha=1}^{\infty}\alpha^{-\lambda}\sum_{\beta_\alpha=1}^{\infty}\beta_\alpha^{-\lambda+1}\right).$$

Since for $\lambda > 2$ we have both

$$\sum_{\beta=1}^{\infty}\beta_\alpha^{-\lambda} = \mathcal{O}(1)$$

and

$$\sum_{\beta=1}^{\infty}\beta_\alpha^{-\lambda+1} = \mathcal{O}(1)$$

we obtain

$$E_{n,k}(x) = \mathcal{O}\left(\frac{1}{k}\right) + \mathcal{O}\left(\frac{1}{n}\right).$$

Now in (13.13) we get

$$|\mathcal{F}_{n,k}(f)(x) - f(x)| \le \left(1 + m\left(\mathcal{O}\left(\frac{1}{k}\right) + \mathcal{O}\left(\frac{1}{n}\right)\right)\right)\omega\left(f, \frac{1}{m}\right).$$

If we put $m = \min\{k, n\}$ we get $m\left(\mathcal{O}\left(\frac{1}{k}\right) + \mathcal{O}\left(\frac{1}{n}\right)\right) = \mathcal{O}(1)$ and finally we obtain that there exists a constant $C$ such that

$$|\mathcal{F}_{n,k}(f)(x) - f(x)| \le C\omega\left(f, \frac{1}{m}\right).$$

∎

**Corollary 13.22.** *If $A_i(x)$ denote Shepard kernels (13.10) having the parameter $\lambda > 2$, then the composition of the inverse and direct F-transform given by (13.3) and (13.2) converges uniformly to $f$.*

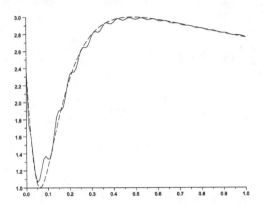

**Fig. 13.9** Approximation by Shepard type F-transform. Dashed line = $f_1$, solid line = inverse F-transform

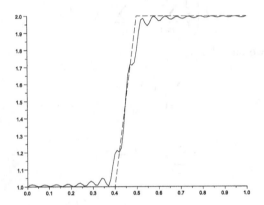

**Fig. 13.10** Approximation by Shepard type F-transform. Dashed line = $f_2$, solid line = inverse F-transform

Similar result can be obtained for the continuous F-transform with Shepard kernels.

**Example 13.23.** *Functions $f_1, f_2$ in (13.8), (13.9) are approximated by Shepard type inverse F-transform operators. The results are shown in Figs. 13.9, 13.10 respectively.*

## 13.5   Korovkin Type Theorems for the F-Transform

The first of Korovkin's Theorems is an important result in classical approximation theory. Let us recall it here.

**Theorem 13.24.** *(Korovkin [96]) Let $L_n \in \mathcal{L}(C[a,b])$, $n = 1, 2, ...$ be a sequence of positive linear operators. If $\lim_{n \to \infty} L_n(e_i) = e_i$, $i = 0, 1, 2$ where $e_0(x) = 1$, $e_1(x) = x, e_2(x) = x^2$, then $\lim_{n \to \infty} L_n(f) = f$, for any $f \in C[a,b]$.*

A particularization can be considered for our F-transforms as follows:

**Theorem 13.25.** *(Korovkin's Theorem for the discrete F-transform, see Bede-Rudas [23]) Let $\mathcal{F}_{n,k} \in \mathcal{L}(C[a,b])$, $n, k = 1, 2, ...$ be the composition between the inverse and direct discrete F-transforms with respect to a fixed fuzzy partition. The following general error estimate holds true for any $\delta > 0$ :*

$$\|\mathcal{F}_{n,k}(f) - f\| \leq \left(1 + \frac{1}{\delta}\sqrt{e_2 - 2e_1\mathcal{F}_{n,k}(e_1) + \mathcal{F}_{n,k}(e_2)}\right)\omega(f,\delta),$$

*where $e_1(x) = x, e_2(x) = x^2$.*

**Proof.** The composition of the inverse and direct discrete F-transform can be expressed as

$$\mathcal{F}_{n,k}(f)(x) = \sum_{i=1}^{k} A_i(x)\frac{\sum_{j=1}^{n} A_i(x_j)f(x_j)}{\sum_{j=1}^{n} A_i(x_j)}.$$

Let us observe that for the inverse F-transform we have $\mathcal{F}_{n,k}(e_0) = e_0$, where $e_0(x) = 1 \ \forall x \in [a,b]$, and it is immediate to observe that

$$|\mathcal{F}_{n,k}(f) - f(x)| \leq \sum_{i=1}^{k} A_i(x)\frac{\sum_{j=1}^{n} A_i(x_j)\,|f(x) - f(x_j)|}{\sum_{j=1}^{n} A_i(x_j)},$$

and then by standard reasoning

$$|\mathcal{F}_{n,k}(f) - f(x)|$$

$$\leq \left(1 + \frac{1}{\delta}\sum_{i=1}^{k} A_i(x)\frac{\sum_{j=1}^{n} A_i(x_j)|x - x_j|}{\sum_{j=1}^{n} A_i(x_j)}\right)\omega(f,\delta),$$

for any fixed $\delta > 0$. The error is controlled by the ratio

$$R_{n,k}(x) = \sum_{i=1}^{k} A_i(x)\frac{\sum_{j=1}^{n} A_i(x_j)|x - x_j|}{\sum_{j=1}^{n} A_i(x_j)}. \tag{13.14}$$

Using Cauchy-Schwarz inequality we have

$$\sum_{j=1}^{n} A_i(x_j)|x - x_i| \leq \left(\sum_{j=1}^{n} A_i(x_j)|x - x_j|^2\right)^{\frac{1}{2}} \left(\sum_{j=1}^{n} A_i(x_j)\right)^{\frac{1}{2}}$$

and using Cauchy-Schwarz inequality successively we obtain

$$R_{n,k}(x) \leq \sum_{i=1}^{k} A_i(x) \frac{\left(\sum_{j=1}^{n} A_i(x_j)|x - x_j|^2\right)^{\frac{1}{2}}}{\left(\sum_{j=1}^{n} A_i(x_j)\right)^{\frac{1}{2}}}$$

$$= \sum_{i=1}^{k} \sqrt{A_i(x)} \frac{\left(A_i(x)\sum_{j=1}^{n} A_i(x_j)|x - x_j|^2\right)^{\frac{1}{2}}}{\left(\sum_{j=1}^{n} A_i(x_j)\right)^{\frac{1}{2}}}$$

$$\leq \left(\sum_{i=1}^{k} A_i(x)\right)^{\frac{1}{2}} \left(\sum_{i=1}^{k} \frac{A_i(x)\sum_{j=1}^{n} A_i(x_j)|x - x_j|^2}{\sum_{j=1}^{n} A_i(x_j)}\right)^{\frac{1}{2}}.$$

Since $A_i(x)$ is a fuzzy partition we have

$$R_{n,k}(x) \leq \left(\sum_{i=1}^{k} A_i(x) \frac{\sum_{j=1}^{n} A_i(x_j)(x^2 - 2xx_j + x_j^2)}{\sum_{j=1}^{n} A_i(x_j)}\right)^{\frac{1}{2}}$$

and finally

$$R_{n,k}(x) = \left(x^2 - 2x\mathcal{F}_{n,k}(e_1)(x) + \mathcal{F}_{n,k}(e_2)(x)\right)^{\frac{1}{2}},$$

which completes the proof.    ∎

**Corollary 13.26.** *If*

$$\lim_{n,k\to\infty} \mathcal{F}_{n,k}(e_i) = e_i,$$

$i = 1, 2$, *then*

$$\lim_{n,k\to\infty} \mathcal{F}_{n,k}(f) = f,$$

*for any $f \in C[a, b]$.*

If we consider the case of the continuous F-transform then we obtain a similar result.

**Theorem 13.27.** *(Korovkin's Theorem for the continuous F-transform, Bede-Rudas [23]) Let $\mathcal{F}_k \in \mathcal{L}(C[a, b])$, $k = 1, 2, ...$ be the composition of the inverse and direct continuous F-transforms. The following general error estimate holds true for any $\delta > 0$:*

$$\|\mathcal{F}_k(f) - f\| \leq \left(1 + \frac{1}{\delta}\sqrt{e_2 - 2e_1\mathcal{F}_k(e_1) + \mathcal{F}_k(e_2)}\right)\omega(f, \delta),$$

*where $e_1(x) = x, e_2(x) = x^2$.*

**Proof.** The proof uses the integral version of Cauchy-Schwarz inequality and it is left to the reader as an exercise. ∎

**Corollary 13.28.** *If* $\lim_{k\to\infty} \mathcal{F}_k(e_i) = e_i$, $i = 1, 2$, *then* $\lim_{n\to\infty} \mathcal{F}_k(f) = f$, *for any* $f \in C[a, b]$.

The Korovkin type Theorems above show that if we are able to have small values of

$$E = \sqrt{e_2 - 2e_1 \mathcal{F}_{n,k}(e_1) + \mathcal{F}_{n,k}(e_2)}$$

for the discrete case, or

$$E = \sqrt{e_2 - 2e_1 \mathcal{F}_k(e_1) + \mathcal{F}_k(e_2)}$$

for the continuous case, for a given sequence of fuzzy partitions $A_i$, $i = 1, ..., k$, then we have small approximation error for any continuous function.

## 13.6   F-Transform with Bernstein Basis Polynomials and the Durrmeyer Operator

Using the Korovkin-type results we can study F-transforms with Bernstein basis polynomials used as a fuzzy partition this time. The proposed construction leads to the well-known Durrmeyer operator (Bede-Rudas [23]). The Bernstein basis polynomials are

$$p_{k,i}(x) = \binom{k}{i} x^i (1 - x)^{k-i},$$

$k = 1, 2, ...$ $i = 0, 1, ..., k$ $x \in [0, 1]$. In Fig. 13.11 elements of Bernstein basis are illustrated.

It is easy to check that

$$\sum_{i=0}^{k} p_{n,i}(x) = (x + (1 - x))^k = 1,$$

i.e. $p_{n,i}$ is a fuzzy partition.

The inverse F-transform is given by

$$\mathcal{F}_k(f)(x) = \sum_{i=0}^{k} p_{k,i}(x) \frac{\int_0^1 p_{k,i}(x) f(x) dx}{\int_0^1 p_{k,i}(x) dx},$$

that is the classical Durrmeyer operator Durrmeyer [53].

**Theorem 13.29.** *Let* $f : [0, 1] \to \mathbb{R}$ *be continuous, and let* $\mathcal{F}_k(f)(x)$ *denote the inverse F-transform approximation operator, i.e., the Durrmeyer operator. Then the following error estimate holds true*

$$\|\mathcal{F}_k(f) - f\| \leq C\omega\left(f, \sqrt{\frac{x(1 - x)}{k}}\right).$$

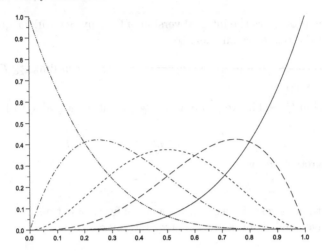

**Fig. 13.11** The Bernstein basis polynomials

**Proof.** Let us calculate $\mathcal{F}_k(e_1)$ and $\mathcal{F}_k(e_2)$ in this case. It is easy to check that

$$p'_{k,i}(x) = k(p_{k-1,i-1}(x) - p_{k-1,i}(x)).$$

By integrating we obtain

$$\int_0^1 p_{k-1,i-1}(x)dx = \int_0^1 p_{k-1,i}(x)dx,$$

i.e., for fixed $k$ the integrals of all Bernstein basis polynomials are the same namely

$$\int_0^1 p_{k,i}(x)dx = \frac{1}{k+1}, i = 1, ..., k.$$

By direct calculation

$$\int_0^1 p_{k,i}(x)x\,dx = \int_0^1 \binom{k}{i} x^{i+1}(1-x)^{k-i}dx$$

$$= \frac{i+1}{k+1} \int_0^1 p_{k+1,i+1}dx = \frac{i+1}{(k+1)(k+2)}.$$

Also,

$$\int_0^1 p_{k,i}(x)x^2\,dx = \int_0^1 \binom{k}{i} x^{i+2}(1-x)^{k-i}dx$$

$$= \frac{i+1}{k+1}\frac{i+2}{k+2} \int_0^1 p_{k+2,i+2}dx$$

$$= \frac{(i+1)(i+2)}{(k+1)(k+2)(k+3)}.$$

Then we obtain

$$\mathbf{f}_i(e_1) = \frac{\int_0^1 p_{k,i}(x)x\,dx}{\int_0^1 p_{k,i}(x)dx} = \frac{i+1}{k+2} = \mathcal{O}\left(\frac{i}{k}\right)$$

and

$$\mathbf{f}_i(e_2) = \frac{\int_0^1 p_{k,i}(x)x^2\,dx}{\int_0^1 p_{k,i}(x)dx}$$

$$= \frac{(i+1)(i+2)}{(k+2)(k+3)} = \mathcal{O}\left(\frac{i^2}{k^2}\right).$$

$$\mathcal{F}_k(e_1)(x) = \mathcal{O}\left(\sum_{i=1}^{k} p_{k,i}(x)\frac{i}{k}\right)$$

$$= \mathcal{O}\left(B_k(e_1)(x)\right) = \mathcal{O}\left(e_1\right),$$

where $B_k$ denotes the usual Bernstein operator

$$B_k(f)(x) = \sum_{i=1}^{k} p_{k,i}(x)f\left(\frac{i}{k}\right)$$

and it is well known that $B_k(e_1)(x) = x$. We also have

$$\mathcal{F}_k(e_2)(x) = \mathcal{O}\left(\sum_{i=1}^{k} p_{k,i}(x)\left(\frac{i}{k}\right)^2\right)$$

$$= \mathcal{O}\left(B_k(e_2)(x)\right) = \mathcal{O}\left(x^2 + \frac{x(1-x)}{k}\right),$$

and finally we obtain: $\sqrt{e_2 - 2e_1 \mathcal{F}_k(e_1) + \mathcal{F}_k(e_2)} = \mathcal{O}\sqrt{\frac{x(1-x)}{k}}$ and by the above Korovkin type results we have

$$\|\mathcal{F}_k(f) - f\| \leq C\omega\left(f, \sqrt{\frac{x(1-x)}{k}}\right).$$

$\blacksquare$

**Example 13.30.** *We consider the problem to approximate $f_1, f_2$ in (13.8), (13.9) by Bernstein inverse F-transform. The results are illustrated in Figs. 13.12, 13.13.*

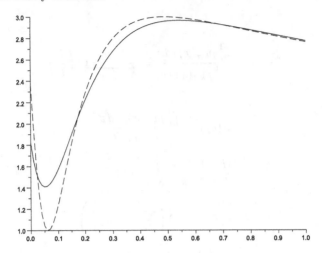

**Fig. 13.12** Approximation by Bernstein type F-trasform. Dashed line = $f_1$ Solid line = inverse F-transform.

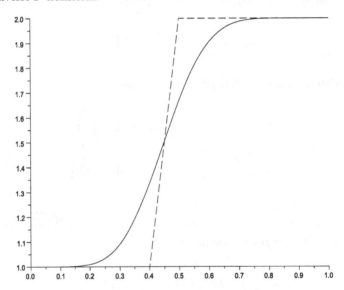

**Fig. 13.13** Approximation by Bernstein type F-transform. Dashed line = $f_2$, solid line = inverse F-transform.

## 13.7  F-Transform of Favard-Szász-Mirakjan Type

It is possible to further generalize Fuzzy Transforms by considering unbounded intervals and infinitely many atoms for a fuzzy partition. Let $f$ : $[0, \infty) \to \mathbb{R}$ be a bounded function. We consider now on the $[0, \infty)$ interval the fuzzy partition $\mathbb{A} = (A_0, A_1, ....)$, with $\sum_{i=0}^{\infty} A_i(x) = 1, \forall x \in I$. We can consider for example a Favard-Szász-Mirakjan kernel (see e.g. Gal [67])

$$A_i(x) = e^{-nx} \frac{(nx)^i}{i!},$$

$n \geq 1, i = 0, 1, ....$ The continuous F-transform in this case is given by

$$f_i = \frac{\int_0^{\infty} A_i(x) f(x) dx}{\int_0^{\infty} A_i(x) dx},$$

and the discrete F-transform is given by

$$f_i = \frac{\sum_{j=1}^{\infty} A_i(x_j) f(x_j)}{\sum_{j=1}^{\infty} A_i(x_j)},$$

where $x_j \in I$, given data ($j \geq 1$). The composition of the continuous inverse and direct F-transform is

$$\mathcal{F}_n(f)(x) = \sum_{i=0}^{\infty} A_i(x) f_i,$$

and in the discrete case

$$\mathcal{F}_{n,k}(f)(x) = \sum_{i=0}^{\infty} A_i(x) f_i,$$

for $x \in [0, \infty)$. We observe here an analogy between fuzzy transform and the Fourier transform, inverse fuzzy transform and Fourier series respectively. We will discuss here the Favard-Szász-Mirakjan type inverse F-transform approximation operator. We restrict $\mathcal{F}_n$ to any compact interval $[0, b] \subset [0, \infty)$.

**Theorem 13.31.** *(Bede-Rudas [23]) Let $f : [0, \infty) \to \mathbb{R}$ be continuous and bounded function and let $\mathcal{F}_n(f)(x)$ denote the continuous inverse F-transform approximation operator. Then the following error estimate*

$$\|\mathcal{F}_n(f) - f\| \leq C\omega \left( f, \sqrt{\frac{x}{n}} \right),$$

*holds true on any compact sub-interval of $[0, \infty)$.*

**Proof.** Then the Korovkin-type theorems can be used. On any fixed compact interval we estimate the quantity

$$\sqrt{e_2 - 2e_1 \mathcal{F}_n(e_1) + \mathcal{F}_n(e_2)}.$$

We have

$$\mathbf{f}_i(e_1) = \frac{\int_0^{\infty} A_i(x) x dx}{\int_0^{\infty} A_i(x) dx} = \frac{\frac{1}{i!n^2} \Gamma(i+2)}{\frac{1}{i!n} \Gamma(i+1)},$$

where $\Gamma(x) = \int_0^\infty t^{x-1}e^{-t}dt$ is Euler's gamma function. It is well known that $\Gamma(n+1) = n!, \forall n \in \mathbb{N}$ and we get

$$\mathbf{f}_i(e_1) = \frac{i+1}{n},$$

Similarly

$$\mathbf{f}_i(e_2) = \frac{(i+1)(i+2)}{n^2}.$$

For the inverse F-transforms we get

$$\mathcal{F}_n(e_1)(x) = \sum_{i=0}^\infty A_i(x)\frac{i+1}{n} = \sum_{i=0}^\infty e^{-nx}\frac{(nx)^i}{i!}\frac{i}{n} + \frac{1}{n}$$

$$= xe^{-nx}\sum_{i=1}^\infty \frac{(nx)^{i-1}}{(i-1)!} + \frac{1}{n} = e_1(x) + \frac{1}{n}.$$

and

$$\mathcal{F}_n(e_2)(x) = \sum_{i=0}^\infty A_i(x)\frac{(i+1)(i+2)}{n^2}$$

$$= \sum_{i=0}^\infty e^{-nx}\frac{(nx)^i}{i!}\frac{i^2}{n^2} + \frac{3x}{n} + \frac{2}{n^2}$$

$$= x^2 e^{-nx}\sum_{i=2}^\infty \frac{(nx)^{i-2}}{(i-2)!} + \frac{4x}{n} + \frac{2}{n^2}$$

$$= x^2 + \frac{4x}{n} + \frac{2}{n^2}.$$

As a conclusion the error is controlled by

$$\sqrt{e_2 - 2e_1 \mathcal{F}_n(e_1) + \mathcal{F}_n(e_2)}(x)$$

$$= \sqrt{x^2 - 2x\left(x + \frac{1}{n}\right) + x^2 + \frac{4x}{n} + \frac{2}{n^2}}$$

$$= \sqrt{\frac{2x}{n} + \frac{2}{n^2}} = \mathcal{O}\left(\sqrt{\frac{x}{n}}\right).$$

∎

**Example 13.32.** *In Figs. 13.14 and 13.15 the approximation capabilities of Favard-Szász-Mirakjan type F-transforms are illustrated.*

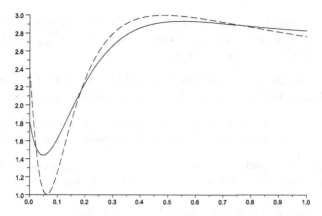

**Fig. 13.14** Approximation by Favard-Szász-Mirakjan type F-trasform. Dashed line $= f_1$, solid line = inverse F-transform.

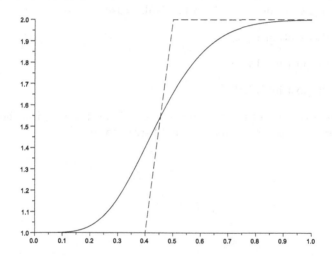

**Fig. 13.15** Approximation by Favard-Szász-Mirakjan type F-transform. Dashed line $= f_2$, solid line = inverse F-transform.

## 13.8 Problems

1. Consider the function $f : [0,1] \to \mathbb{R}$, $f(x) = x$ and let $x_i = \frac{i}{n}$, $i = 0, ..., n$. For a fuzzy partition with triangular fuzzy numbers $A_i(x) = (x_{i-1}, x_i, x_{i+1})$, $i = 1, ..., n-1$ calculate the F-transforms $f_i$, $i = 1, ..., n$ and then calculate the inverse F-transform $F(x) = \sum_{i=1}^{n} A_i(x) \cdot f_i$. Perform detailed calculations for $n = 4$.

2. Consider the function $f : [0,1] \to \mathbb{R}$, $f(x) = x^2$ and let $x_i = \frac{i}{n}$, $i = 0, ..., n$. For a fuzzy partition with triangular fuzzy numbers $A_i(x)$

$= (x_{i-1}, x_i, x_{i+1})$, $i = 1, ..., n-1$ calculate the F-transforms $f_i$, $i = 1, ..., n$ and then calculate the inverse F-transform $F(x) = \sum_{i=1}^{n} A_i(x) \cdot f_i$. Perform detailed calculations for $n = 4$.

3. Using the results of the previous two problems show that the error term in the Korovkin type Theorem 13.27 converges to 0 as $n \to \infty$.

4. Consider the function $f : [0, 1] \to \mathbb{R}$, $f(x) = e^x$. Calculate the direct and inverse F-transform for a triangular fuzzy partition for the given function.

5. Let us consider equally spaces knots $y_i = \frac{i}{n}$, $i = 0, ..., n$ in the $[0, 1]$ interval . Let us consider $y_{-1} \le 0$, $1 \le y_{n+1}$ two auxiliary knots. Let us take

$$A_i(x) = \begin{cases} \frac{1}{2}\left(\cos\frac{\pi}{h}(x - y_i) + 1\right) & \text{if} \quad x \in [y_{i-1}, y_{i+1}] \\ 0 & \text{otherwise} \end{cases} \quad i = 0, ..., n.$$

Calculate the F-transform and the inverse F-transform of the function $f(x) = e^x$. Consider $n = 4$ as a particular case.

6. Prove Proposition 13.8.

7. Prove Proposition 13.10.

8. Prove Proposition 13.12.

9. Starting from one of the F-transform techniques presented in the present chapter construct a two-dimensional F-transform.

# 14

# Artificial Neural Networks and Neuro-Fuzzy Systems

## 14.1 Artificial Neuron

Computational Intelligence is a discipline within Artificial Intelligence and it studies the topics of Fuzzy Sets and Systems, Neural Networks, Genetic Algorithms, Swarm Intelligence and combination of these topics (see Engelbrecht [55]). The present chapter will present an introduction to the Theory of Neural Networks, and also the combination of Neural and Fuzzy systems, i.e., the Adaptive Network-based Fuzzy Inference System (Jang [81]).

A neural network is a nonlinear mapping $f : \mathbb{R}^n \to \mathbb{R}^m$, where $m, n$ are the dimensions of the input and output spaces. The building block of a neural network is an artificial neuron, or simply neuron. The simplest neural network has its input signals $x_1, ..., x_n$ transformed into an output $y$, by propagating them through the synapses, with synaptic weights $w_1, ..., w_n$ with a bias $b$ and an activation function $\varphi$. The structure is described as in Fig. 14.1.

The definition of a neural network in this simple form is as follows:

**Definition 14.1.** *A neural network is a function of the form*

$$y = \varphi \left( \sum_{j=1}^{n} w_j x_j + b \right),$$

*where $\varphi : \mathbb{R} \to [a, b]$ is an activation function, i.e., $\lim_{x \to -\infty} \varphi(x) = a$ and $\lim_{x \to \infty} \varphi(x) = b$.*

B. Bede: *Mathematics of Fuzzy Sets and Fuzzy Logic*, STUDFUZZ 295, pp. 247–258.
DOI: 10.1007/978-3-642-35221-8_14     © Springer-Verlag Berlin Heidelberg 2013

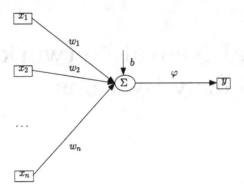

**Fig. 14.1** The structure of an artificial neuron

Different types of activation functions can be used in a neuron, as for example

- Threshold function

$$\varphi(x) = \begin{cases} 1 & \text{if} & x > 0 \\ 0 & , & \text{otherwise} \end{cases} .$$

- Piecewise linear

$$\varphi(x) = \begin{cases} 0 & \text{if} & x < -\varepsilon \\ \frac{1}{2\varepsilon}(x + \varepsilon) & \text{if} & -\varepsilon \le x \le \varepsilon \\ 1 & \text{if} & \varepsilon < x \end{cases} .$$

- Sigmoid

$$\varphi(x) = \frac{1}{1 + e^{-x}}.$$

- tanh

$$\varphi(x) = \tanh x.$$

- etc.

## 14.2    Feed-Forward Neural Network

The simple structure in the previous section can be generalized by adding different layers with different entities. Feed-forward neural network architecture has three layers: Input layer, hidden layer and output layer. The input layer contains the input variables of the system. The hidden layer has

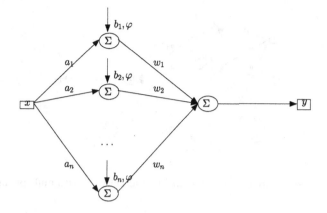

**Fig. 14.2** The structure of a feed-forward neural network with a single input

multiple neurons. The output layer is aggregating the output that comes from the intermediate hidden layer with given synaptic weights. Figure 14.2 illustrates the structure of a feed-forward neural network with a single input.

**Definition 14.2.** *A feed forward neural network is a function*

$$y = \sum_{j=1}^{n} w_j \varphi \left( a_j x + b_j \right),$$

*where* $\varphi : \mathbb{R} \to [a, b]$ *is an activation function.*

A more general case is that of a network handling multiple inputs. The inputs are linked to the hidden layer through different synaptic weights. In Fig. 14.3, a feed-forward neural network with multiple inputs is shown.

**Definition 14.3.** *A feed-forward neural network with multiple inputs and single output, is a function*

$$y = \sum_{j=1}^{n} w_j \varphi \left( \sum_{i=1}^{m} a_{ij} x_i + b_j \right),$$

*where* $\varphi : \mathbb{R} \to [a, b]$ *is an activation function.*

We may have neural networks with multidimensional input and output (see Fig. 14.4).

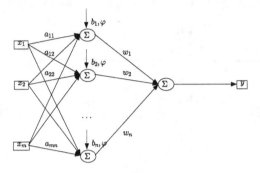

**Fig. 14.3** The structure of a feed-forward neural network with multiple inputs and a single output

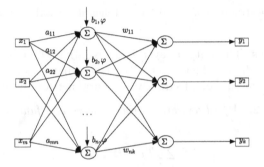

**Fig. 14.4** The structure of a feed-forward neural network with multiple inputs and a single output

**Definition 14.4.** *A feed-forward neural network with both multiple inputs $x_i, i = 1, ..., m$ and outputs $y_l, l = 1, ..., k$ is expressed as a function*

$$y_l = \sum_{j=1}^{n} w_{lj}\varphi\left(\sum_{i=1}^{m} a_{ij}x_i + b_j\right), l = 1, ..., k$$

*where $\varphi : \mathbb{R} \to [a, b]$ is an activation function.*

**Remark 14.5.** *Neural networks are motivated by several important properties:*

- *Nonlinearity. Since $\varphi$ is a nonlinear function, it can approximate well nonlinear phenomena.*

- *Input output mapping. Typical inputs of the neural network should be mapped into typical outputs.*

- *Learning. The pairs of corresponding input-output values provided in a given application are used for training a neural network so that it fulfills the input-output mapping property. The neural network weights are updated according to a learning rule, to fulfill these conditions.*

- *Adaptivity. A neural network is able to adapt when the input-output values change.*

- *Fault Tolerance. A neural network is a robust system. If one single neuron is lost the network output does not change significantly.*

## 14.3   Learning of a Neural Network

Learning of a neural network is the process of updating the synaptic weights such that given known inputs, propagated through the network, lead to approximations of the desired outputs. The learning process can be supervised, unsupervised, reinforcement learning.

Supervised learning provides input-output data for the network, and the network is adapting its weights such that the input-output mapping is realized. In an unsupervised learning no input-output data is provided for the network, and the network is adapting its weights such that a clustering of data is realized. Reinforcement learning updates the synaptic weights in such a way that the neurons that produce the worst output are punished (reinforced). For different learning algorithms see e.g. Engelbrecht [55], Thrun [148].

Let us consider a feed-forward neural network with one hidden layer of the particular form

$$y = \sum_{j=1}^{n} w_j \varphi \left( x - x_j \right).$$

A set of training inputs $\hat{x}_i, i = 1, ..., k$, is considered and then the corresponding training (typical) outputs are $\hat{y}_i, i = 1, ..., k$. The actual output is

$$o_i = \sum_{j=1}^{n} w_j \varphi \left( \hat{x}_i - x_j \right).$$

Then the error function

$$E = \sum_{i=1}^{k} (o_i - \hat{y}_i)^2 = \sum_{i=1}^{k} \left( \sum_{j=1}^{n} w_j \varphi \left( \hat{x}_i - x_j \right) - \hat{y}_i \right)^2.$$

We have

$$\frac{\partial E}{\partial w_j} = 2 \sum_{i=1}^{k} (o_i - \hat{y}_i) \frac{\partial o_i}{\partial w_j} = 2 \sum_{i=1}^{k} (o_i - \hat{y}_i) \varphi \left( \hat{x}_i - x_j \right).$$

The basic algorithm for the learning process of a neural network is the gradient descent algorithm. It is described as follows:

- Step 1. Assume random synaptic weights $w_j, j = 1, ..., k$ and $\varepsilon > 0$.

- Step 2. Calculate the actual outputs given the training data,

$$o_i = \sum_{j=1}^{n} w_j \varphi \left( \hat{x}_i - x_j \right)$$

- Step 3. Update the synaptic weights

$$w_i^{iter+1} = w_i^{iter} - \eta \frac{\partial E}{\partial w_j^{iter}}$$

- Step 4. If the error

$$E = \sum_{i=1}^{k} (o_i - \hat{y}_i)^2 > \varepsilon$$

and if a maximal number of iterations is not attained then go to step 2.

Gradient descent algorithm can be used as the learning algorithm for the synaptic weights in the hidden layer. Let

$$y = \sum_{j=1}^{n} w_j \varphi \left( \sum_{i=1}^{m} a_{ij} x_i + b_j \right),$$

be a feed-forward network. We assume given training data $\hat{x}_{li}, \hat{y}_l, l = 1, ..., k$, $i = 1, ..., m$. The error function is

$$E = \sum_{l=1}^{k} (o_l - \hat{y}_l)^2$$

$$= \sum_{l=1}^{k} \left( \sum_{j=1}^{n} w_j \varphi \left( \sum_{i=1}^{m} a_{ij} \hat{x}_{li} - x_j \right) - \hat{y}_l \right)^2.$$

We will find the gradient. We have

$$\frac{\partial E}{\partial a_{ij}} = 2 \sum_{l=1}^{k} (o_l - \hat{y}_l) \frac{\partial o_l}{\partial a_{ij}},$$

$$\frac{\partial E}{\partial a_{ij}} = 2\sum_{i=1}^{k}(o_l - \hat{y}_l)w_j\frac{d\varphi}{du}\hat{x}_{li},$$

where $\frac{d\varphi}{du}$ can be calculated based on the activation function. If $\varphi$ is the sigmoid function then we have

$$\frac{\partial E}{\partial a_{ij}} = 2\sum_{i=1}^{k}(o_l - \hat{y}_l)o_l(1 - o_l)\hat{x}_{li}.$$

This is based on the fact that for the sigmoid function $\varphi$ we have $\varphi'(u) = \varphi(u)(1 - \varphi(u))$.

If $\varphi$ is the threshold function then this partial derivative is always 0. This is not a convenient result, so in this case we need a different learning rule. In this case the derivative $\frac{d\varphi}{du}$ is replaced by the approximation $o_l(1 - o_l)$. So we have the same expression

$$\frac{\partial E}{\partial a_{ij}} = 2\sum_{i=1}^{k}(o_l - \hat{y}_l)o_l(1 - o_l)\hat{x}_{li}.$$

## 14.4   Approximation Properties of Neural Networks

We denote by $C[0,1]$ the space of continuous functions endowed with the uniform norm. We will show that any continuous function can be uniformly approximated by neural networks using a constructive approach. Let $0 = x_0 \le x_1 \le ... \le x_n \le x_{n+1} = 1$ be a partition with equally spaces points and $f \in C[0,1]$. We define the neural network

$$N_n f(x) = \sum_{i=0}^{n}\omega_i \tau(x - x_i),$$

where $\tau$ is the threshold function and

$$\omega_i = \frac{\chi(f,i)}{2} - \sum_{j=0}^{i-1}\omega_j,$$

with

$$\chi(f,i) = \sup_{x\in[x_i,x_{i+1})} f(x) + \inf_{x\in[x_i,x_{i+1})} f(x), \ i = 0,...,n$$

**Theorem 14.6.** *(see Csáji [40])The following error estimate holds true*

$$\|f - N_n f\| \le \frac{\omega(f,\frac{1}{n})}{2}.$$

**Proof.** As the first step of the proof we will show that $N_n f(x) = \frac{\chi(f,i)}{2}$. Indeed, if $x \in [x_i, x_{i+1})$ then

$$N_n f(x) = \sum_{i=0}^{n} \omega_i \tau(x - x_i) = \sum_{j=0}^{i-1} \omega_j + \omega_i$$

$$= \sum_{j=0}^{i-1} \omega_j + \frac{\chi(f,i)}{2} - \sum_{j=0}^{i-1} \omega_j = \frac{\chi(f,i)}{2}.$$

Now we turn our attention to the estimate. First we observe that

$$\omega\left(f, \frac{1}{n}\right) \geq \sup_{x,y \in [x_i, x_{i+1})} |f(x) - f(y)| = \sup_{x \in [x_i, x_{i+1})} f(x) - \inf_{x \in [x_i, x_{i+1})} f(x)$$

$$= \sup_{x \in [x_i, x_{i+1})} f(x) - \frac{\chi(f,i)}{2} + \frac{\chi(f,i)}{2} - \inf_{x \in [x_i, x_{i+1})} f(x)$$

$$= 2 \sup_{x \in [x_i, x_{i+1})} \left| f(x) - \frac{\chi(f,i)}{2} \right|.$$

Now, for $x \in [x_i, x_{i+1})$

$$\sup_{x \in [x_i, x_{i+1})} |f(x) - N_n f(x)| = \sup_{x \in [x_i, x_{i+1})} |f(x) - \frac{\chi(f,i)}{2}| \leq \frac{1}{2} \omega\left(f, \frac{1}{n}\right)$$

and as a conclusion we get

$$\|f - N_n f\| \leq \frac{\omega(f, \frac{1}{n})}{2}.$$

∎

**Corollary 14.7.** *If $f$ is continuous then $\lim_{n \to \infty} N_n f(x) = f(x)$.*

The theorem can be extended to the case of sigmoid activation function (Cybenko [41]) and other neural networks (Anastassiou [3]).

**Example 14.8.** *Let us consider the function $f_1 = e^{-\frac{x}{5}} \sin 2x$. Its neural network approximation is given in Fig. 14.5.*

## 14.5   Adaptive Network Based Fuzzy Inference System (ANFIS)

An adaptive network based fuzzy inference system (ANFIS Jang [81]) is able to combine a fuzzy system's ability to model a reasoning process and to handle uncertainty, with the learning ability and adaptivity of a neural network. ANFIS is based on a Takagi-Sugeno fuzzy system. Let us consider an example of a Takagi Sugeno fuzzy system with a single input and output and $n$ fuzzy rules.

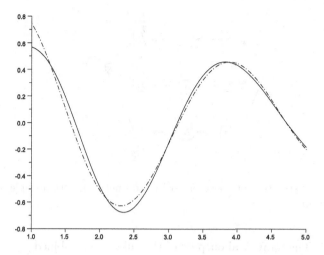

**Fig. 14.5** Approximation by a neural network. Dashed line $f_1$ and solid line =neural network

- Rule 1: If $x$ is $A_1$ then $f_1 = p_1 x + r_1$

- ......

- Rule $n$: If $x$ is $A_n$ then $f_n = p_n x + r_n$

The architecture of the ANFIS is as in Fig. 14.6.
  The layers are described as follows:

- The input layer receives data $x$.

- Layer 1: The membership grades $A_1(x), A_2(x), ..., A_n(x)$ are calculated.

- Layer 2: The firing levels of the fuzzy rules, namely

$$w_1 = A_1(x) \ ... \ w_n = A_n(x)$$

  are calculated.

- Layer 3: Normalized firing levels

$$w_1' = \frac{w_1}{w_1 + ... + w_n} \text{ and } w_n' = \frac{w_n}{w_1 + ... + w_n}$$

  are calculated.

- Layer 4: The individual output of each fuzzy rule can now be calculated
  as

$$w_1' f_1 = w_1'(p_1 x + r_1),$$

$$...$$

$$w_n' f_n = w_n'(p_n x + r_n).$$

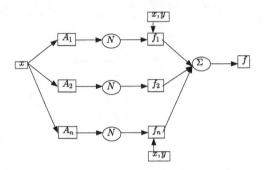

**Fig. 14.6** The structure of a feedforward neural network with a single inputs and a single output

- Layer 5: The individual outputs of the rules are combined

$$f = \sum_{i=1}^{n} w_i'(p_i x + r_i).$$

- The output layer has outcome $f$.

The difference between ANFIS and a Takagi-Sugeno system is that the weights, and the membership functions in ANFIS are learned through an optimization algorithm. Surely, as Takagi-Sugeno systems without an added optimization are approximation operators, the ANFIS system will be able to approximate any continuous function.

Let us consider an example of a Takagi Sugeno fuzzy system with two linguistic antecedents, and two fuzzy rules.

- Rule 1: If $x$ is $A_1$ and $y$ is $B_1$ then $f_1 = p_1 x + q_1 y + r_1$

- Rule 2: If $x$ is $A_2$ and $y$ is $B_2$ then $f_2 = p_2 x + q_2 y + r_2$

The architecture of the ANFIS is as in Fig. 14.7.

The layers are described as follows:

- The input layer receives data $x, y$.

- Layer 1: The membership grades $A_1(x), A_2(x)$ and $B_1(x), B_2(x)$ are calculated. ($A_i, B_i$, are given fuzzy sets).

- Layer 2: The firing levels of the fuzzy rules, namely

$$w_1 = A_1(x) \cdot B_1(y) \text{ and } w_2 = A_2(x) \cdot B_2(y)$$

are calculated.

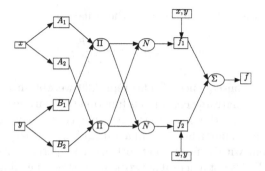

**Fig. 14.7** The structure of a feedforward neural network with multiple inputs and a single output

- Layer 3: Normalized firing levels

$$w_1' = \frac{w_1}{w_1 + w_2} \text{ and } w_2' = \frac{w_2}{w_1 + w_2}$$

are calculated.

- Layer 4: The individual output of each fuzzy rule can now be calculated as

$$w_1' f_1 = w_1'(p_1 x + q_1 y + r_1)$$

and

$$w_2' f_2 = w_2'(p_2 x + q_2 y + r_2)$$

- Layer 5: The individual outputs of the two rules are combined

$$f = w_1' f_1 + w_2' f_2 = w_1'(p_1 x + q_1 y + r_1) + w_2'(p_2 x + q_2 y + r_2)$$

- The output layer has outcome $f$.

The learning of an ANFIS can adaptively derive the coefficient values $p_1, p_2$, $q_1, q_2, r_1, r_2$ using a least squares method. Also, parameters of the fuzzy sets $A_1, A_2, B_1, B_2$ can be determined or adjusted by a suitable learning algorithm. This improves the flexibility and adaptability of the network.

Let us describe for example the learning algorithm for the parameter $p_1$. The Error function can be written

$$E = \sum_{j=1}^{n}(f(x_j, y_j) - z_j)^2,$$

where $x_j, y_j, z_j, j = 1, ..., n$ is training data. Then

$$\frac{\partial E}{\partial p_1} = 2\sum_{j=1}^{n}(f(x_j, y_j) - z_j)\frac{\partial f}{\partial p_1} = 2\sum_{j=1}^{n}(f(x_j, y_j) - z_j)w_1' x_j.$$

The gradient descent algorithm is then the usual one

$$p_1^{iter+1} = p_1^{iter} - \eta \frac{\partial E}{\partial p_1^{iter}}.$$

As an immediate consequence of Theorem 7.30 we obtain that the ANFIS system can approximate any continuous function with any accuracy, moreover the order of approximation is at least the same as in Theorem 7.30. Indeed, ANFIS is in essence a particular case of a Takagi-Sugeno fuzzy system. ANFIS is a fuzzy system with a neural network used to perform its optimization adaptation. ANFIS is not a neural network in the strict meaning of Definitions 14.1 or 14.3. Surely the idea of adaptivity combined with that of fuzziness makes the system very useful and extremely powerful in applications.

## 14.6   Problems

1. Consider the Neural Network

$$y_l = \sum_{j=1}^{n} w_{lj} \varphi \left( \sum_{i=1}^{m} a_{ij} x_i + b_j \right),$$

with $\varphi(x) = \frac{1}{1+e^{\alpha x}}$, $\alpha > 0$ being a parameter. Deduce learning rules for the synaptic weights $w_{lj}, a_{ij}$.

2. Repeat the previous problem with the activation function $\varphi(x) = \tanh x$.

3. Consider

$$\varphi(x) = \begin{cases} -1 & \text{if} & x < -\frac{\pi}{2} \\ \sin x & \text{if} & -\frac{\pi}{2} \le x \le \frac{\pi}{2} \\ 1 & \text{if} & \frac{\pi}{2} < x \end{cases}.$$

Show that $\varphi$ is an activation function and use this function to construct a neural network with one hidden layer. Describe learning rules for the given network.

4. Given $\varphi(x)$ an activation function prove that $\varphi(ax)$ with $a > 0$ is also an activation function. Use this method to construct neural networks with variants of tanh and sin as given above.

5. Prove that ANFIS can approximate any continuous function based on Theorem 7.30,

# References

[1] Aczél, J.: Lectures on functional equations and their applications. Academic Press (1966)

[2] Alsina, C., Trillas, E., Valverde, L.: On Some Logical Connectives for Fuzzy Sets Theory. Journal of Mathematical Analysis and its Applications 93, 15–26 (1983)

[3] Anastassiou, G.A.: Rate of Convergence of some neural network operators to the unit-univariate case. Journal of Mathematical Analysis and Applications 212, 237–262 (1997)

[4] Anastassiou, G.A.: Fuzzy Mathematics: Approximation Theory. Springer, Heidelberg (2010)

[5] Anastassiou, G.A., Gal, S.G.: On a fuzzy trigonometric approximation theorem of Weierstrass type. Journal of Fuzzy Mathematics 9, 701–708 (2001)

[6] Atanassov, K.: Intuitionistic fuzzy sets. Fuzzy Sets and Systems 20, 87–96 (1986)

[7] Ban, A.I.: Intuitionistic Fuzzy Measures: Theory and Applications. Nova Science Publishers (2006)

[8] Ban, A.I., Bede, B.: Cross product of L-R fuzzy numbers and applications. Annals of the University of Oradea, Fasc. Math. 9, 95–108 (2003)

[9] Ban, A.I., Bede, B.: Properties of the cross product of fuzzy numbers. Journal of Fuzzy Mathematics 14, 513–531 (2006)

[10] Ban, A.I., Bede, B.: Power series of fuzzy numbers with cross product and applications to fuzzy differential equations. Journal of Concrete and Applicable Mathematics 4, 125–152 (2006)

[11] Ban, A.I., Coroianu, L.C., Grzegorzewski, P.: Trapezoidal approximation and aggregation. Fuzzy Sets and Systems 177, 45–59 (2011)

[12] Barros, L.C., Bassanezi, R.C., Tonelli, P.A.: On the continuity of the Zadeh's extension. In: Proc. IFSA 1997 Congress, Prague (1997)

[13] Barros, L.C., Bassanezi, R.C.: Tópicos de Lógica Fuzzy e Biomatemática. In: Unicamp/Imecc (2010) (in Portuguese)

[14] Bede, B.: A note on "Two-point boundary value problems associated with non-linear fuzzy differential equations". Fuzzy Sets and Systems 157, 986–989 (2006)

[15] Bede, B.: Note on "Numerical solutions of fuzzy differential equations by predictor–corrector method". Information Sciences 178, 1917–1922 (2008)

[16] Bede, B., Coroianu, L., Gal, S.G.: Approximation and Shape Preserving Properties of the Bernstein Operator of Max-Product Kind. International Journal of Mathematics and Mathematical Sciences 2009, Article ID 590589 (2009), doi:10.1155/2009/590589

[17] Bede, B., Fodor, J.: Product Type Operations between Fuzzy Numbers and their Applications in Geology. Acta Polytechnica Hungarica 2, 123–139 (2006)

[18] Bede, B., Gal, S.G.: Almost periodic fuzzy-number-valued functions. Fuzzy Sets and Systems 147, 385–403 (2004)

[19] Bede, B., Gal, S.G.: Quadrature Rules for Fuzzy-Number-Valued Functions. Fuzzy Sets and Systems 145, 359–380 (2004)

[20] Bede, B., Gal, S.G.: Generalizations of the differentiability of fuzzy-number-valued functions with applications to fuzzy differential equations. Fuzzy Sets and Systems 151, 581–599 (2005)

[21] Bede, B., Gal, S.G.: Solutions of fuzzy differential equations based on generalized differentiability. Communications in Mathematical Analysis 9, 22–41 (2010)

[22] Bede, B., Nobuhara, H., Dankova, M., Di Nola, A.: Approximation by using pseudo-linear operators. Fuzzy Sets and Systems 159, 804–820 (2008)

[23] Bede, B., Rudas, I.J.: Approximation properties of fuzzy transforms. Fuzzy Sets and Systems 180, 20–40 (2011)

[24] Bede, B., Rudas, I.J., Bencsik, A.: First order linear differential equations under generalized differentiability. Information Sciences 177, 1648–1662 (2007)

[25] Bede, B., Rudas, I.J., Gal, S.G.: Almost periodic solutions of fuzzy differential equations. In: Proceedings Internat. Sympos. on Computational Intelligence and Intelligent Informatics, ISCIII 2005, Tunis, Tunisia, October 14-16, pp. 58–61 (2005)

[26] Bede, B., Tenali, G.B., Lakshmikantham, V.: Perspectives of Fuzzy Initial Value Problems. Communications in Applied Analysis 11, 339–358 (2007)

[27] Bede, B., Stefanini, L.: Generalized Differentiability of Fuzzy-valued Functions. Fuzzy Sets and Systems (to appear), http://ideas.repec.org/p/urb/wpaper/12_09.html

[28] Bezdek, J.C.: Pattern Recognition with Fuzzy Objective Function Algorithms. Plenum Press, New York (1981)

[29] Bica, A.M.: One-sided fuzzy numbers and applications to integral equations from epidemiology. Fuzzy Sets and Systems (to appear)

[30] Buckley, J.J., Feuring, T.: Fuzzy differential equations. Fuzzy Sets and Systems 110, 43–54 (2000)

[31] Buckley, J.J., Qu, Y.: On using $\alpha$-cuts to evaluate fuzzy equations. Fuzzy Sets and Systems 38, 309–312 (1990)

[32] Buckley, J.J., Jowers, L.: Simulating Continuous Fuzzy Systems. Springer, Heidelberg (2006)

[33] Buckley, J.J.: Sugeno type controllers are universal approximators. Fuzzy Sets and Systems 53, 299–303 (1993)

[34] Celikyilmaz, A., Türksen, I.B.: Modeling Uncertainty with Fuzzy Logic: With Recent Theory and Applications. Springer (2009)

[35] Chalco-Cano, Y., Román-Flores, H., Jiménez-Gamero, M.D.: Generalized derivative and $\pi$-derivative for set-valued functions. Information Sciences 181, 2177–2188 (2011)

[36] Chalco-Cano, Y., Román-Flores, H.: On new solutions of fuzzy differential equations Chaos. Solitons & Fractals 38, 112–119 (2008)

[37] Chalco-Cano, Y., Román-Flores, H.: Comparation between some approaches to solve fuzzy differential equations. Fuzzy Sets and Systems 160, 1517–1527 (2009)

[38] Cignoli, R.L.O., D'Ottaviano, I.M.L., Mundici, D.: Algebraic Foundations of Many-valued Reasoning. Kluwer, Dordrecht (2000)

[39] Coman, G., Tambulea, L.: A Shepard-Taylor approximation formula. Studia Univ. Babes-Bolyai Math. 33, 65–73 (1988)

[40] Csáji, B. C.: Approximation with Artificial Neural Networks, MSc thesis Eötvös Loránd University Hungary (2001)

[41] Cybenko, G.: Approximations by superpositions of sigmoidal functions. Mathematics of Control, Signals, and Systems 2, 303–314 (1989)

[42] De Baets, B.: Analytical Solution Methods for Fuzzy Relational Equations. In: Dubois, D., Prade, H. (eds.) Fundamentals of Fuzzy Sets. Kluver (2000)

[43] DeVore, R.A., Lorentz, G.G.: Constructive Approximation. Polynomials and Splines Approximation. Springer, Heidelberg (1993)

[44] Diamond, P., Kloeden, P.: Metric topology of fuzzy numbers and fuzzy analysis. In: Dubois, D., Prade, H., et al. (eds.) Handbook Fuzzy Sets Ser., vol. 7, pp. 583–641. Kluwer Academic Publishers, Dordrecht (2000)

[45] Diamond, P.: Stability and periodicity in fuzzy differential equations. IEEE Transactions on Fuzzy Systems 8, 583–590 (2000)

[46] Ding, Z., Ma, M., Kandel, A.: Existence of the solutions of fuzzy differential equations with parameters. Information Sciences 99, 205–217 (1997)

[47] Di Nola, A., Lettieri, A.: Equational characterization of all varieties of MV-algebras. Journal of Algebra 221, 123–131 (1993)

[48] Di Nola, A., Sessa, S., Pedrycz, W., Sanchez, E.: Fuzzy Relation Equations and Their Applications to Knowledge Engineering. Springer (1989)

[49] Dubois, D., Prade, H.: Fuzzy Sets and Systems, Theory and Applications. Academic Press (1980)

[50] Dubois, D., Prade, H.: Fuzzy numbers: An overview. In: Analysis of Fuzzy Information. Mathematical Logic, vol. 1, pp. 3–39. CRC Press, Boca Raton (1987)

[51] Dubois, D., Prade, H.: Fundamentals of fuzzy sets. Kluver Academic Publishers (2000)

[52] Dubois, D., Martin-Clouaire, R., Prade, H.: Practical computation in fuzzy logic. In: Gupta, M.M., Yamakawa, T. (eds.) Fuzzy Computing, pp. 11–34. Elsevier Science Publishing, Amsterdam (1988)

[53] Durrmeyer, J.L.: Une formule d'inversion de la transformáe de Laplace: Applications á la théorie des moments. Thése de 3e cycle, Paris (1967)

[54] Dvorak, A., Novak, W.: A Fuzzy Logic Model of Detective Reasoning, University of Ostrava, Institute for Research and Applications of Fuzzy Modeling, Research Report No. 99 (2006)

[55] Engelbrecht, A.P.: Computational Intelligence, an Introduction, 2nd edn. John Wiley & Sons (2007)

[56] Erceg, M.A.: Metric Spaces in Fuzzy Set Theory. Journal of Mathematical Analysis and its Applications 69, 205–230 (1979)

[57] Faber, V.: Clustering and the continuousk-means algorithm. Los Alamos Sci. 22, 138–144 (1994)

[58] Fodor, J.: On fuzzy implication operators. Fuzzy Sets and Systems 42, 293–300 (1991)

[59] Fodor, J.: Left-continuous t-norms in fuzzy logic: An overview. Acta Polytechnica Hungarica 1 (2004)

[60] Fodor, J., Roubens, M.: Fuzzy Preference Modeling and Multicriteria Decision Support. Kluwer (1994)

[61] Fodor, J., Yager, R.R.: Fuzzy Set Theoretic Operations and Quantifiers. In: Dubois, D., Prade, H. (eds.) Fundamentals of Fuzzy Sets. Kluwer (2000)

[62] Fortin, J., Dubois, D., Fargier, H.: Gradual Numbers and their Application to Fuzzy Interval Analysis. IEEE Transactions on Fuzzy Systems 16, 388–402 (2008)

[63] Fullér, R.: Neural Fuzzy Systems. Abo Akademi University (1995)

[64] Fullér, R.: Introduction to Neuro-Fuzzy Systems. Springer (2000)

[65] Fullér, R., Keresztfalvi, T.: On generalization of Nguyen's theorem. Fuzzy Sets and Systems 41, 371–374 (1990)

[66] Gal, S.G.: Approximation Theory in Fuzzy Setting. In: Anastassiou, G.A. (ed.) Handbook of Analytic-Computational Methods in Applied Mathematics, ch. 13, pp. 617–666. Chapman & Hall/CRC, Boca Raton (2000)

[67] Gal, S.G.: Global Smoothness and Shape Preserving Interpolation by Classical Operators. Birkhäuser (2005)

[68] Gal, S.G.: Shape-Preserving Approximation by Real and Complex Polynomials. Birkhäuser, Boston (2008)

[69] Gal, S.G.: Linear continuous functionals on FN-type spaces. J. Fuzzy Math. 17(3), 535–553 (2009)

[70] Gal, S.G., Ban, A.I.: Elemente de Matematica Fuzzy. University of Oradea (1996) (in Romanian)

[71] Gal, C.S., Gal, S.G.: Semigroups of operators on spaces of fuzzy-number-valued functions with applications to fuzzy differential equations. J. Fuzzy Math. 13, 647–682 (2005)

[72] Gal, C.S., Gal, S.G., N'guerekata, G.M.: Existence and uniqueness of almost automorphic mild solutions to some semilinear fuzzy differential equations, Trends in African diaspora math. research, pp. 23–35. Nova Sci. Publ., Huntington (2007)

[73] Goguen, J.A.: L-fuzzy sets. J. Math. Anal. Appl. 18, 145–174 (1967)

[74] Goetschel, R., Voxman, W.: Elementary fuzzy calculus. Fuzzy Sets and Systems 18, 31–43 (1986)

[75] Grzegorzewski, P., Mrówka, E.: Trapezoidal approximations of fuzzy numbers. Fuzzy Sets and Systems 153(1), 115–135 (2005)

[76] Hajek, P.: Metamathematics of Fuzzy Logic. Kluwer (1998)

[77] Hanss, M.: Applied Fuzzy Arithmetic. Springer (2005)

[78] Huang, H.: Some notes on Zadeh's extensions. Information Sciences 180, 3806–3813 (2010)

[79] Hukuhara, M.: Integration des applications measurables dont la valeur est un compact convexe. Funkcialaj Ekvacioj 10, 205–223 (1967)

[80] Hullermeier, E.: An approach to modeling and simulation of uncertain dynamical systems. Int. J. Uncertainty Fuzziness Knowledge-Based Syst. 5, 117–137 (1997)

[81]  Jang, J.S.R.: ANFIS: Adaptive-Network-Based Fuzzy Inference System. IEEE
      Trans. Systems, Man, Cybernetics 23, 665–685 (1993)
[82]  Jain, A.K., Dubes, R.C.: Algorithms for Clustering Data. Prentice Hall (1988)
[83]  Kaleva, O.: Fuzzy differential equations. Fuzzy Sets and Systems 24, 301–317
      (1987)
[84]  Kaleva, O.: A note on fuzzy differential equations. Nonlinear Analysis 64,
      895–900 (2006)
[85]  Kaleva, O.: Nonlinear iteration semigroups of fuzzy Cauchy problems. Fuzzy
      Sets and Systems (to appear)
[86]  Khastan, A., Nieto, J.J.: A boundary value problem for second order fuzzy
      differential equations. Original Research Article Nonlinear Analysis: Theory,
      Methods & Applications 72, 3583–3593 (2010)
[87]  Klement, E.P., Mesiar, R., Pap, E.: Triangular Norms. Kluwer, Dordrecht
      (2000)
[88]  Klement, E.P., Mesiar, R., Pap, E.: Triangular norms. Position paper I: basic
      analytical and algebraic properties. Fuzzy Sets and Systems 143(1), 5–26
      (2004)
[89]  Klement, E.P., Mesiar, R., Pap, E.: Triangular norms. Position paper II: gen-
      eral constructions and parameterized families. Fuzzy Sets and Systems 145(3),
      411–438 (2004)
[90]  Klement, E.P., Mesiar, R., Pap, E.: Triangular norms. Position paper III:
      continuous t-norms. Fuzzy Sets and Systems 145(3), 439–454 (2004)
[91]  Klir, G.J.: Fuzzy arithmetic with requisite constraints. Fuzzy Sets and Sys-
      tems 91, 165–175 (1997)
[92]  Klir, G.J.: The role of constrained fuzzy arithmetic in engineering. In: Ayyub,
      B.M. (ed.) Uncertainty Analysis in Engineering and the Sciences. Kluwer,
      Boston (1997)
[93]  Klir, G.J., Pan, Y.: Constrained fuzzy arithmetic: Basic questions and some
      answers. Soft Computing 2, 100–108 (1998)
[94]  Klir, G.J., Yuan, B.: Fuzzy sets and fuzzy logic: theory and applications.
      Prentice Hall PTR (1995)
[95]  Kóczy, L.T., Zorat, A.: Fuzzy systems and approximation. Fuzzy Sets and
      Systems 85, 203–222 (1997)
[96]  Korovkin, P.P.: On convergence of linear and positive operators in the space
      of continuous functions. Dokl. Akad. Nauk. SSSR (N.S.) 90, 961–964 (1953)
      (Russian)
[97]  Kosko, B.: Fuzzy Systems as Universal Approximators. IEEE Transactions
      on Computers 43, 1329–1333 (1994)
[98]  Lakshmikantham, V., Mohapatra, R.N.: Theory of Fuzzy Differential Equa-
      tions and Inclusions. CRC Press (2003)
[99]  Lee, K.H.: First Course on Fuzzy Theory and Applications. Advances in In-
      telligent and Soft Computing series, vol. 27. Springer (2005)
[100] Li, Y.-M., Shi, Z.-K., Li, Z.-H.: Approximation theory of fuzzy systems based
      upon genuine many-valued implications-SISO cases. Fuzzy Sets and Sys-
      tems 130, 147–157 (2002)
[101] Li, Y.-M., Shi, Z.-K., Li, Z.-H.: Approximation theory of fuzzy systems based
      upon genuine many-valued implications-MIMO cases. Fuzzy Sets and Sys-
      tems 130, 159–174 (2002)
[102] Lucyna, R., Karolina, T.: Approximation by modified Favard operators. Com-
      mentationes Mathematicae 44, 205–215 (2004)

[103] Lupulescu, V.: On a class of fuzzy functional differential equations. Fuzzy Sets and Systems 160, 1547–1562 (2009)

[104] Ma, M.: On embedding problem of fuzzy number space part IV. Fuzzy Sets and Systems 58, 185–193 (1993)

[105] MacQueen, J.B.: Some Methods for classification and Analysis of Multivariate Observations. In: Proceedings of 5th Berkeley Symposium on Mathematical Statistics and Probability, pp. 281–297. University of California Press (1967)

[106] Malinowski, M.T.: Interval Cauchy problem with a second type. Hukuhara derivative Information Sciences 213, 94–105 (2012)

[107] Mamdani, E.H., Assilian, S.: An experiment in linguistic synthesis with a fuzzy logic controller. J. Man Machine Stud. 7, 1–13 (1975)

[108] Markov, S.: Extended interval arithmetic. Compt. Rend. Acad. Bulg. Sci. 30, 1239–1242 (1977)

[109] Markov, S.: Calculus for interval functions of a real variable. Computing 22, 325–377 (1979)

[110] Mendel, J.M.: Advances in type-2 fuzzy sets and systems. Information Sciences 177, 84–110 (2007)

[111] Mitaim, S., Kosko, B.: The shape of fuzzy sets in adaptive function approximation. IEEE Transactions on Fuzzy Systems 9, 637–656 (2001)

[112] Miyakoshi, M., Shimbo, M.: Solutions of Composite Fuzzy Relational Equations with Triangular Norms. Fuzzy Sets and Systems 16, 53–63 (1985)

[113] Murofushi, T., Sugeno, M.: An interpretation of fuzzy measures and the Choquet integral as an integral with respect to a fuzzy measure. Fuzzy Sets and Systems 29, 201–227 (1989)

[114] Navara, M., Zabortsky, Z.: How to make constrained fuzzy arithmetic efficient. Soft Computing 5, 412–417 (2001)

[115] Negoita, C., Ralescu, D.: Application of Fuzzy Sets to System Analysis. Wiley, New York (1975)

[116] Nguyen, H.T.: A note on the extension principle for fuzzy sets. J. Math. Anal. Appl. 64, 369–380 (1978)

[117] Nieto, J.J., Rodriguez-Lopez, R.: Analysis of a logistic differential model with uncertainty. International Journal of Dynamical Systems and Differential Equations 1, 164–176 (2008)

[118] Nobuhara, H., Bede, B., Hirota, K.: On Various Eigen Fuzzy Sets and Their Application to Image Reconstruction. Information Sciences 176, 2988–3010 (2006)

[119] Pedrycz, W., Gomide, F.: An Introduction to Fuzzy Sets, Analysis and Design. MIT Press (1998)

[120] Perfilieva, I.: Fuzzy Transforms. In: Peters, J.F., Skowron, A., Dubois, D., Grzymała-Busse, J.W., Inuiguchi, M., Polkowski, L. (eds.) Transactions on Rough Sets II. LNCS, vol. 3135, pp. 63–81. Springer, Heidelberg (2004)

[121] Perfilieva, I.: Fuzzy Transforms: theory and applications. Fuzzy Sets and Systems 157, 993–1023 (2006)

[122] Perko, L.: Differential Equations and Dynamical Systems. Springer (2000)

[123] Puri, L.M., Ralescu, D.: Differentials of fuzzy functions. J. Math. Anal. Appl. 91, 552–558 (1983)

[124] Ralescu, D., Adams, G.: The fuzzy integral. Journal of Mathematical Analysis and Applications 75, 562–570 (1980)

[125] Rodríguez-López, R.: Comparison results for fuzzy differential equations. Information Sciences 178, 1756–1779 (2008)

[126] Rojas-Medar, M., Roman-Flores, H.: On the equivalence of convergences of fuzzy sets. Fuzzy Sets and Systems 80, 217–224 (1996)

[127] Roman-Flores, H., Barros, L.C., Bassanezi, R.C.: A note on Zadeh's extensions. Fuzzy Sets and Systems 117, 327–331 (2001)

[128] Rudas, I.J., Pap, E., Fodor, J.: Information aggregation in intelligent systems: An application oriented approach. Knowledge-Based Systems (to appear, 2012)

[129] Sanchez, E.: Resolution of composite fuzzy relation equations. Information and Control 30, 38–48 (1976)

[130] Schweizer, B., Sklar, A.: Probabilistic Metric Spaces. North-Holland, Amsterdam (1983)

[131] Seikkala, S.: On the fuzzy initial value problem. Fuzzy Sets and Systems 24, 319–330 (1987)

[132] Sonbol, A.H., Fadali, M.S.: TSK Fuzzy Function Approximators: Design and Accuracy Analysis. IEEE Transactions on Systems Man and Cybernetics 42, 702–712 (2012)

[133] Song, S., Wu, C.: Existence and uniqueness of solutions to the Cauchy problem of fuzzy differential equations. Fuzzy Sets and Systems 110, 55–67 (2000)

[134] Stancu, D.D., Coman, G., Blaga, P.: Analiza numerica si teoria aproximarii, vol. II. University Press, Cluj-Napoca (2002)

[135] Stefanini, L.: On the generalized LU-fuzzy derivative and fuzzy differential equations. In: Proceedings of the 2007 IEEE International Conference on Fuzzy Systems, London, pp. 710–715 (July 2007)

[136] Stefanini, L.: A generalization of Hukuhara difference and division for interval and fuzzy arithmetic. Fuzzy Sets and Systems 161, 1564–1584 (2010)

[137] Stefanini, L., Sorini, L., Guerra, M.L.: Parametric representation of fuzzy numbers and application to fuzzy calculus. Fuzzy Sets and Systems 157, 2423–2455 (2006)

[138] Sugeno, M.: An introductory survey of fuzzy control. Information Sciences 36, 59–83 (1985)

[139] Sugeno, M., Murofushi, T.: Pseudo-additive measures and integrals. Journal of Mathematical Analysis and Applications 122, 197–222 (1987)

[140] Sussner, P., Nachtegael, M., Mélange, T., Deschrijver, G., Esmi, E., Kerre, E.: Interval-Valued and Intuitionistic Fuzzy Mathematical Morphologies as Special Cases of L -Fuzzy Mathematical Morphology. Journal of Mathematical Imaging and Vision 43, 50–71 (2012)

[141] Stefanini, L., Bede, B.: Generalized Hukuhara differentiability of interval-valued functions and interval differential equations. Nonlinear Analysis: Theory, Methods & Applications 71, 1311–1328 (2009)

[142] Szabados, J.: Direct and converse approximation theorems for the Shepard operator. Approximation Theory and Applications 7, 63–76 (1991)

[143] Szmidt, E., Kacprzyk, J.: Distances between intuitionistic fuzzy sets. Fuzzy Sets and Systems 114, 505–518 (2000)

[144] Tarski, A.: A lattice-theoretical fixpoint theorem and its applications. Pacific Journal of Mathematics 5, 285–309 (1955)

[145] Tenali, G.B., Lakshmikantham, V., Devi, V.: Revisiting fuzzy differential equations. Nonlinear Analysis 58, 351–358 (2004)

[146] Tikk, D., Kóczy, L.T., Gedeon, T.D.: A survey on universal approximation and its limits in soft computing techniques. International Journal of Approximate Reasoning 33, 185–202 (2003)

[147] Trillas, E.: Sobre funciones de negacion en la teoria de conjuntos difusos. Stochastica 3, 47–60 (1979)

[148] Thrun, S.: Explanation-Based Neural Network Learning: A Lifelong Learning Approach. Kluwer (1996)

[149] Wagenknecht, M., Hampel, R., Schneider, V.: Computational aspects of fuzzy arithmetics based on Archimedean t-norms. Fuzzy Sets and Systems 123, 49–62 (2001)

[150] Wu, C., Song, S., Stanley Lee, E.: Approximate solutions, existence and uniqueness of the Cauchy problem of fuzzy differential equations. Journal of Mathematical Analysis and Applications 202, 629–644 (1996)

[151] Wu, C., Gong, Z.: On Henstock integral of fuzzy-number-valued functions I. Fuzzy Sets and Systems 120, 523–532 (2001)

[152] Xiao, W., Zhou, S.P.: A Jackson type estimate for Shepard operators in $L^p$ spaces for $p \geq 1$. Acta Matheamtica Hungarica 95, 217–224 (2002)

[153] Ying, H.: General Takagi-Sugeno fuzzy systems with simplified linear rule consequent are universal controllers, models and filters. Information Sciences 108, 91–107 (1998)

[154] Zadeh, L.A.: Fuzzy Sets. Information and Control 8, 338–353 (1965)

[155] Zadeh, L.A.: Outline of a new approach to the analysis of complex systems and decision processes. IEEE Transactions on Systems, Man and Cybernetics 3, 28–44 (1973)

[156] Zadeh, L.A.: The concept of a linguistic variable and its application to approximate reasoning. Information Sciences 8, 199–249 (1975)

[157] Zadeh, L.A.: Computing with Words. STUDFUZZ, vol. 277. Springer, Heidelberg (2012)

# Appendix A
# Mathematical Prerequisites

In the present Appendix we recall a few basic definitions and theorems from Mathematics, especially Mathematical Analysis, that are used throughout the text. This is not intended to be an exhaustive treatise or a replacement of a solid mathematics background for students, instead we will concentrate on results cited in the text in an effort to make the text more independent.

## A.1  Lattices

Let $L$ be a set and $\leq$ be a relation (classical relation) on $L$.

**Definition A.1.** *A classical relation is said to be a partial order on $L$ if it is:*

  *(i) reflexive, i.e., $a \leq a, \forall a \in L$*
  *(ii) antisymmetric, i.e., if $a \leq b$ and $b \leq a$ then $a = b$, $a, b \in L$*
  *(iii) transitive i.e., if $a \leq b$ and $b \leq c$ then $a \leq c$, $a, b, c \in L$*
  *We say that $L$ is a partially ordered set (poset).*

**Example A.2.** *As examples of partial orders we mention: $\leq$ the standard inequality of real numbers, $\subseteq$ inclusion of classical sets or fuzzy sets, $|$ divisibility of integers, etc.*

Let us recall the following general definition of the infimum and supremum.

**Definition A.3.** *(i) The greatest lower bound (infimum) of two elements $a, b$ of a poset $L$ is an element $c \in L$ such that $c \leq a$, $c \leq b$ and for any $c' \leq a$ and $c' \leq b$ we have $c' \leq c$. We denote $c = \inf\{a, b\}$ or $c = a \wedge_L b$.*

*(ii) The least upper bound (supremum) of two elements $a, b$ of a poset $L$ is an element $d \in L$ such that $a \leq d$, $b \leq d$ and for any $a \leq d'$ and $b \leq d'$ we have $d \leq d'$. We denote $d = \sup\{a, b\}$ or $c = a \vee_L b$.*

*(iii) The greatest lower bound (infimum) of a nonempty subset $A \subseteq L$ is $c \in L$ if $c \leq x, \forall x \in A$ and if $c' \in L$ is such that $c' \leq x, \forall x \in A$ then $c' \leq c$.*

*(iv) The least upper bound (supremum) of a nonempty subset $A \subseteq L$ is $d \in L$ if $x \leq d, \forall x \in A$ and if $d' \in L$ is such that $x \leq d', \forall x \in A$ then $d \leq d'$.*

One of the possible ways to define a lattice is the following.

**Definition A.4.** *We say that a partially ordered set $L$ is a lattice if any two elements in $L$ have a supremum and an infimum.*

A lattice can also be defined based on interpretation of the inf and sup as algebraic operations.

**Example A.5.** *$(\mathbb{R}, \leq)$, $(\mathbb{N}, |)$, $(\mathcal{P}(X), \subseteq)$ are lattices. In $(\mathbb{R}, \leq)$ the infimum is the minimum and supremum coincides with the maximum. In $(\mathbb{N}, |)$ the infimum of two elements $a, b$ is their greatest common divisor, while their supremum is their least common multiple.*

If the inf and sup operations are distributive w.r.t. each-other then we call the lattice a distributive lattice.

**Definition A.6.** *A complete lattice is a partially ordered set $(L, \leq)$, such that any non-empty subset of $L$ has an infimum and a supremum.*

**Definition A.7.** *A lattice $L$ is called completely distributive if*

$$a \wedge \left( \bigvee_{i \in I} b_i \right) = \bigvee_{i \in I} a \wedge b_i$$

*and*

$$a \vee \left( \bigwedge_{i \in I} b_i \right) = \bigwedge_{i \in I} a \vee b_i$$

*hold $\forall a, b_i \in L, i \in I$ and $I$ any index set.*

**Example A.8.** *$([0, 1], \leq)$ is a complete, completely distributive lattice.*

The next theorems are due to Tarski–Knaster and Kleene respectively (see e.g. Tarski [144])

**Theorem A.9.** *Let $L$ be a complete lattice and $f : L \to L$ be a monotonic function. If the set $\{x \in L | f(x) \leq x\}$ ($\{x \in L | x \leq f(x)\}$) is non-empty, then the set of fixed points forms a complete lattice and $f$ has a least (greatest) fixed point $x_* \in L$ ($x^* \in L$).*

**Theorem A.10.** *Let $L$ be a complete lattice having first and last element that is $\exists 0, 1 \in L$, i.e. $0 \leq x, \forall x \in L$ and $x \leq 1, \forall x \in L$) and let $f : L \to L$ be $\leq$-continuous (i.e., $f$ is such that for $c_1 \leq c_2 \leq \ldots \leq c_n \leq \ldots$ we have*

$$\bigvee_{n=1}^{\infty} f(c_n) = f\left(\bigvee_{n=1}^{\infty} c_n\right).$$

*Then the least and greatest fixed points satisfy*

$$x_* = \bigvee_{n=1}^{\infty} f^n(0), \quad x^* = \bigwedge_{n=1}^{\infty} f^n(1),$$

*where $f^n$ means here $n$ times composition of $f$ by $f$.*

## A.2 Real Numbers

**Definition A.11.** *We consider a field $\mathbb{R}$ (i.e., a set endowed with additive and multiplicative operation with certain properties), with a total order relation (reflexive, transitive, antisymmetric relation, such that any two elements are comparable) compatible with the operations. If additionally $\mathbb{R}$ has the least upper bound property then $\mathbb{R}$ is called the real numbers set, and its elements are called real numbers.*

The least upper bound property is an axiom of the real-numbers and it is used throughout the text.

**Axiom A.12.** *(Least Upper Bound property) Any nonempty set $A \subseteq \mathbb{R}$ bounded from above (i.e., if there exists $M \in \mathbb{R}$ with $x \leq M, \forall x \in A$), has a supremum (least upper bound).*

Dually any nonempty set bounded from below (i.e., if there exists $m \in \mathbb{R}$ with $m \leq x, \forall x \in A$) will have an infimum (greatest lower bound).

## A.3 Metric Spaces

Let us recall here the definition of a metric space.

**Definition A.13.** *Let $X$ be a set and $d : X \times X \to [0, \infty)$ be a function. The function $d$ is called a metric on $X$ (and in this case $X$ is called a metric space) if the following properties are fulfilled.*
  *(i) $d(x, y) \geq 0, \forall x, y \in X$ with $d(x, y) = 0$ if and only if $x = y$.*
  *(ii) $d(x, y) = d(y, x), \forall x, y \in X$.*
  *(iii) $d(x, z) \leq d(x, y + d(y, z), \forall x, y, z \in X$.*

As examples of metric spaces let us mention $\mathbb{R}^n$ with e.g. Euclidean distance which in the one-dimensional case becomes $d(x, y) = |x - y|$.

**Definition A.14.** *A sequence $(x_n)_{n\geq 0} \subseteq X$ is said to converge to $x \in X$ if $\forall \varepsilon > 0$ there exist $N(\varepsilon) \geq 0$ such that $d(x_n, x) < \varepsilon, \forall n \geq N(\varepsilon)$.*

**Definition A.15.** *A sequence $(x_n)_{n\geq 0} \subseteq I$ is fundamental if $\forall \varepsilon > 0$ there exist $N(\varepsilon) \geq 0$ such that $d(x_n, x_{n+p}) < \varepsilon, \forall n \geq N(\varepsilon), p \geq 1$.*

**Definition A.16.** *A metric space is complete if any fundamental sequence converges.*

The Euclidean space $\mathbb{R}^n$ and in particular $\mathbb{R}$ are complete metric spaces.

**Definition A.17.** *An open ball centered at $x_0$ and radius $r$ in a metric space $X$ is*

$$B(x_0, r) = \{x \in X | d(x, x_0) < r\}.$$

*A closed ball in a metric space is defined as*

$$\bar{B}(x_0, r) = \{x \in X | d(x, x_0) \leq r\}.$$

*A set $V \subseteq X$ is a neighborhood of $x \in X$ if it contains an open ball $x \in B(x, r) \subseteq V$. A set $A \subseteq X$ is open if it is a neighborhood of every of its points.*

An important property is that arbitrary unions of open sets are open.

**Definition A.18.** *A set is closed if its complement is open.*

Sets may be closed and open simultaneously, so the concepts of open and closed set do not exclude each-other.

**Proposition A.19.** *A set $A \subseteq X$ is closed if and only if any convergent sequence $(x_n)_{n\geq 0} \subseteq A$ converges to $x \in A$.*

**Definition A.20.** *The closure of a set is the least closed set (in the sense of the inclusion) that contains the set.*

The closure of a set consists of all the points of the set and the limits of all sequences with terms in the set. For example the closure of the set of rational numbers $\mathbb{Q}$ is the set of real numbers $\mathbb{R}$. The notation $cl(A)$ is used to denote the closure of the set $A$.

**Definition A.21.** *A subset $A \subseteq X$ is said to be compact if from any covering $A \subseteq \bigcup_{i \in I} U_i$ with open sets we can extract a finite sub-covering $A \subseteq \bigcup_{i=1}^n U_i$.*

A compact set must be closed. A subset of a metric space is compact if and only if it is sequentially compact (i.e, any bounded sequence has a convergent subsequence). Also, a subset $A$ of a complete metric space is compact if and only if it is totally bounded, i.e. $\forall r > 0$ we can cover $A$ with finitely many open balls $B(x_i, r), i = 1, .., n$ with the radius $r$.

**Theorem A.22.** *(Heine-Borel) A subset $A \subseteq \mathbb{R}^n$ is compact if and only if it is closed and bounded.*

A Lindelöf space is a generalization of the concept of compactness.

**Definition A.23.** *A subset $A \subseteq X$ is said to be Lindelöf if from any covering $A \subseteq \bigcup_{i \in I} U_i$ with open sets we can extract a countable sub-covering $A \subseteq \bigcup_{i=1}^{\infty} U_i$.*

Separability is another important property.

**Definition A.24.** *A metric space $X$ is said to be separable if it is the closure of a countable subset $A \subseteq X$.*

If a space is compact then it is also separable. For example $\mathbb{R}$ is separable being the closure of $\mathbb{Q}$.

Another important property is convexity.

**Definition A.25.** *A subset $A \subseteq \mathbb{R}^n$ is said to be convex if for any $x, y \in A$ we have $\lambda x + (1 - \lambda)y \in A$.*

A subset of $\mathbb{R}$ is a closed interval if and only if it is both compact and convex.

## A.4   Continuity

**Definition A.26.** *A function $f : X \to Y$ is continuous at $x \in X$ if for any neighborhood $V$ of $f(x)$ we can find a neighborhood $U$ of $x$ with $f(U) \subseteq V$.*

Equivalently we can formulate continuity using the distances on $X$ and $Y$.

**Proposition A.27.** *A function $f : X \to Y$ is continuous at $x \in X$ if for any $\varepsilon > 0$ there exists $\delta > 0$ such that $d_2(f(y), f(x)) < \varepsilon$ whenever $d_1(x, y) < \delta$. If it is continuous for every value of $x \in X$ then we say it is continuous on $X$. (We denote the distance on $X$ by $d_1$ and the distance on $Y$ by $d_2$.)*

Continuity and compactness work together in the following way.

**Proposition A.28.** *The image of a compact set $K \subseteq X$ through a continuous function $f : X \to Y$ ($X, Y$ metric spaces) is compact.*

The space of all continuous functions $f : K \to Y$ on a compact metric space $K$, with $Y$ being a metric space, is denoted by $C(K, Y)$. On $C(K, Y)$ we can define the metric

$$D(f, g) = \sup_{x \in K} d(f(x), f(g)),$$

and then the function space itself will become a metric space. The metric $D$ on $C(K, Y)$ is called the uniform distance. If additionally $Y$ is complete then $C(K, Y)$ is a complete metric space with respect to the uniform distance.

Upper semicontinuity is used when we define fuzzy numbers.

**Definition A.29.** *A function $f : X \to \mathbb{R}$ ($X$ being a metric space) is upper semicontinuous at $x_0$ if $\forall \varepsilon > 0$ there exists $\delta > 0$ such that $f(x) < f(x_0) + \varepsilon$ for $d(x, y) < \delta$.*

**Theorem A.30.** *A function is upper semicontinuous if and only if the sets*

$$\{x \in X | f(x) < \alpha\}, \alpha \in \mathbb{R}$$

*are open, or equivalently if the sets*

$$\{x \in X | f(x) \geq \alpha\}, \alpha \in \mathbb{R}$$

*are closed.*

**Definition A.31.** *A function $f : X \to \mathbb{R}$ (X being a metric space) is lower semicontinuous at $x_0$ if $\forall \varepsilon > 0$ there exists $\delta > 0$ such that $f(x) > f(x_0) - \varepsilon$ for $d(x, y) < \delta$.*

**Theorem A.32.** *A function is lower semicontinuous if and only if the sets*

$$\{x \in X | f(x) \leq \alpha\}, \alpha \in \mathbb{R}$$

*are closed, or equivalently if the sets*

$$\{x \in X | f(x) > \alpha\}, \alpha \in \mathbb{R}$$

*are open.*

If a function is both upper and lower semicontinuous then it is continuous.

Left and right continuity appear in a characterization theorem.

**Definition A.33.** *A function $f : \mathbb{R} \to \mathbb{R}$ is right continuous at $x_0$ if $\forall \varepsilon > 0$ there exists $\delta > 0$ such that $|f(x) - f(x_0)| < \varepsilon$ for $x_0 < x < x_0 + \delta$.*

**Definition A.34.** *A function $f : \mathbb{R} \to \mathbb{R}$ is left continuous at $x_0$ if $\forall \varepsilon > 0$ there exists $\delta > 0$ such that $|f(x) - f(x_0)| < \varepsilon$ for $x_0 - \delta < x < x_0$.*

If a function is both left and right continuous then it is continuous.

**Definition A.35.** *Let $(X, d_1)$, $(Y, d_2)$ be metric spaces. A function $f : X \to Y$ is of Lipschitz type if there exists $L$ such that $d_2(f(x), f(y)) \leq L d_1(x, y)$, $\forall x, y \in X$.*

**Definition A.36.** *Let $(X, d)$ be a metric space. A function $f : X \to X$ is a contraction if there exists $0 < q < 1$ such that $d(f(x), f(y)) \leq q d(x, y)$, $\forall x, y \in X$.*

**Theorem A.37.** *(Banach fixed point theorem) Let $(X, d)$ be a complete metric space. Any contraction $f : X \to X$ has a unique fixed point, i.e., there exists a unique $x_* \in X$ such that $f(x_*) = x_*$.*

## A.5   Modulus of Continuity

The modulus of continuity is often used in the text so, let us recall its definition and main properties.

**Definition A.38.** *Let $(X, d)$ be a compact metric space and $([0, \infty), |\cdot|)$ the metric space of positive reals endowed with the usual Euclidean distance. Let $f : X \to [0, \infty)$ be bounded. Then the function*

$$\omega(f, \cdot) : [0, \infty) \to [0, \infty),$$

*defined by*

$$\omega(f, \delta) = \bigvee \{|f(x) - f(y)|; x, y \in X, \ d(x, y) \leq \delta\}$$

*is called the modulus of continuity of $f$.*

**Theorem A.39.** *The following properties hold true*
  *i)*

$$|f(x) - f(y)| \leq \omega(f, d(x, y))$$

*for any $x, y \in X$;*
  *ii) $\omega(f, \delta)$ is nondecreasing in $\delta$;*
  *iii) $\omega(f, 0) = 0$;*
  *iv)*

$$\omega(f, \delta_1 + \delta_2) \leq \omega(f, \delta_1) + \omega(f, \delta_2)$$

*for any $\delta_1, \delta_2, \in [0, \infty)$;*
  *v)*

$$\omega(f, n\delta) \leq n\omega(f, \delta)$$

*for any $\delta \in [0, \infty)$ and $n \in \mathbb{N}$;*
  *vi)*

$$\omega(f, \lambda\delta) \leq (\lambda + 1) \cdot \omega(f, \delta)$$

*for any $\delta, \lambda \in [0, \infty)$;*
  *vii) $f$ is continuous if and only if*

$$\lim_{\delta \to 0} \omega(f, \delta) = 0.$$

## A.6   Normed Spaces

**Definition A.40.** *Consider a linear space $X$. A norm is a function $\|\cdot\| : X \to [0, \infty)$ having the following properties.*
  *1) $\|x\| \geq 0, \forall x \in X$ and $\|x\| = 0$ if and only if $x = 0$.*
  *2) $\|\lambda x\| = |\lambda| \|x\|, \forall \lambda \in \mathbb{R}, x \in X$.*
  *3) $\|x + y\| \leq \|x\| + \|y\|, \forall x, y \in X$.*
  *In this case the space $X$ is called a normed linear space or simply a normed space.*

The norm induces on $X$ a metric structure with the distance being $d(x, y) = \|x - y\|$.

**Definition A.41.** *If a normed space is a complete metric space with the induced metric, then it is called a Banach space.*

We denote by $C([a, b])$ the space of continuous functions $f : [a, b] \to \mathbb{R}$ with the uniform norm $\|f\| = \sup_{x \in [a,b]} |f(x)|$. Then $C([a, b])$ is a Banch space.

Also, let $\mathcal{L}(C([a, b]))$ be the space of linear operators $T : C([a, b]) \to C([a, b])$. Inequality between functions is considered point-wise, i.e., $f \leq g$ if $f(x) \leq g(x)$ for any $x \in [a, b]$. In particular we say that $f \geq 0$ if $f(x) \geq 0$ for every $x \in [a, b]$.

An operator is said to be positive if $T(f) \geq 0$ whenever $f \geq 0$.

# Index